Measuring the Value of Invention

The Impact of Lemelson-MIT Prize Winners' Inventions

BENJAMIN M. MILLER, DAVID METZ, JON SCHMID, PAIGE M. RUDIN, MARJORY S. BLUMENTHAL

Sponsored by the Lemelson Foundation

RAND SOCIAL AND ECONOMIC WELL-BEING

For more information on this publication, visit www.rand.org/t/RRA838-1

Library of Congress Cataloging-in-Publication Data is available for this publication.
ISBN: 978-1-9774-0654-5

Cover photos: The Lemelson-MIT Program

Support RAND
Make a tax-deductible charitable contribution at
www.rand.org/giving/contribute

www.rand.org

Preface

For the past 25 years, the Lemelson–Massachusetts Institute of Technology (MIT) Program has given an annual $500,000 prize to a mid-career inventor whose work offers a significant value to society, has improved lives and communities, and has been adopted or has a high probability of being adopted for practical use. The Lemelson-MIT Program and the Lemelson Foundation asked the RAND Corporation to provide a careful and impartial assessment of the value provided to society by the inventions of recipients of the Lemelson-MIT Prize. This report provides that assessment, considering impact in aggregate by all prize winners and impact through individual case studies of prize winners from three particular years. The work is intended to be of interest to groups that engage in diverse efforts to support invention and innovation, policymakers seeking guidance on the benefits of supporting invention, and the general reader.

RAND Social and Economic Well-Being is a division of the RAND Corporation that seeks to actively improve the health and social and economic well-being of populations and communities throughout the world. This research was conducted in the Community Health and Environmental Policy Program within RAND Social and Economic Well-Being. The program focuses on such topics as infrastructure, science and technology, community design, community health promotion, migration and population dynamics, transportation, energy, and climate and the environment, as well as other policy concerns that are influenced by the natural and built environment, technology, and community organizations and institutions that affect well-being. For more information, email chep@rand.org.

Contents

A separate appendix (RR-A838-2, *Measuring the Value of Invention: The Impact of Lemelson-MIT Prize Winners' Inventions—Appendix*) is available at www.rand.org/t/RRA838-1.html.

Figures

Tables

Summary

Inventions, such as new tools, devices, processes, and medicines, have provided significant benefits to society. Inventions help people around the world live longer, healthier, and more-productive lives and provide new ways to build, move, communicate, heal, learn, and play. Understanding and clearly communicating the value of invention can help policymakers understand the benefits of policies, programs, and initiatives that support the development of inventions and that address inequities that suppress the development of female and minority inventors.

This report illustrates the value of invention, using the impacts of the inventions of the winners of the Lemelson–Massachusetts Institute of Technology (MIT) Prize as examples. We consider both aggregate impacts of the collective inventions of all prize winners and the impacts of individual prize winners' inventions from three particular years.

Inventions provide a wide variety of benefits; this report documents economic, social, scientific, and technical impacts. The economic and social impacts of prize winners are reflected in the more than 180 organizations affiliated with the 26 Lemelson-MIT Prize winners and their inventions. Two companies founded by Lemelson-MIT Prize winners were among the first biotechnology firms to achieve a market capitalization exceeding $100 billion. Several other publicly traded firms founded by Lemelson-MIT Prize winners had market valuations between $100 million and $50 billion as of 2020. As of 2019, 35 other companies founded by prize winners had been acquired by or had merged with other business entities in market transactions valued at approximately $7.5 billion (in 2019 dollars).[1] Of those companies that remained independent entities as of 2019, a small number grew into large profitable companies, while the vast majority remained successful smaller companies or were start-ups attempting to commercialize new ideas or technologies. Several of these companies did not succeed. Those independent companies that report financial data collectively employ approximately 40,000 workers and generate total annual revenues exceeding $54 billion.[2]

The combined technologies, resources, and capabilities of these organizations have achieved significant advances in several areas, including pharmaceutical manufacturing, drug delivery systems, pattern recognition–based computer technology, mobility devices, and energy storage systems. In some cases, the inventions of the Lemelson-MIT Prize winners not only provided the scientific and technological foundation for new companies but were revolutionary ideas that

---

[1] In addition, 22 more of these companies were involved in acquisitions for which our research team was not able to find publicly available financial details.

[2] Financial data are not available for all companies, including many privately held companies. Several start-up companies reported $0 in annual revenues.

spawned entirely new industries. This broader scientific and technical impact is reflected in the citation of their work in scientific publications, new product patents, licensing, or through open-source technologies. As of March 2020, the 26 prize winners had published more than 3,700 articles that had accumulated more than 334,000 citations in the Web of Science citation index. As of January 1, 2020, their 3,871 patents had been cited as prior art by more than 40,000 subsequent patents.

In addition to these impacts, we highlight three lessons that can be drawn from this assessment. First, the impacts of many of the inventions go beyond the direct effect of the invention. These inventions are often platforms or proofs of concept that enable or inspire entirely new fields of study or industries. For example, the recombinant DNA (deoxyribonucleic acid) technology developed by Herbert Boyer and Stanley Cohen was licensed to hundreds of companies, resulting in thousands of new products in the emerging field of biotechnology. By 2020, the U.S. biotechnology industry had grown to approximately 2,200 firms, 288,000 employees, and $108 billion in annual revenues (IBISWorld, 2020).

Second, successful inventors help solve major challenges facing society. For example, the inventions described in this report's case studies are continuing to help address a wide variety of diseases. The role of inventors in addressing global health challenges continues to play out today, as inventors help tackle challenges created by the current severe acute respiratory syndrome coronavirus 2 (SARS-COV-2, otherwise known as coronavirus disease 2019 [COVID-19]) pandemic. For example, several of the companies founded by Lemelson-MIT Prize winners have efforts underway to address challenges related to the pandemic, including blood testing for antibodies, convalescent plasma therapy for COVID-19 patients, and technology to support public health initiatives such as contact tracing (Lee and Grant, 2020; PathCheck Foundation, undated; Vuturo, 2020).

Third, maximizing the impact of an invention requires more than just creating an invention. Addressing challenges in applying and manufacturing an invention, as well as navigating regulatory or business-management hurdles, requires a talented team and (in some cases) a bit of luck. This also highlights the fact that challenges in supporting invention remain. Evidence suggests that a lack of exposure to role models and support networks is limiting diversity in the population of inventors. Programs and policies that support the development of both inventions and inventors remain important for ensuring everyone has the opportunity to develop as an inventor, as the benefits of future inventions will be greatest in a society that welcomes diverse perspectives and brings all creative minds to bear on the world's challenges.

Acknowledgments

This report has benefited greatly from the insights, support, and feedback of many individuals and organizations throughout this research process. We thank the Massachusetts Institute of Technology's Betsy Boyle, Wendy Brown, Michael Cima, Stephanie Couch, David Gordon, Stephanie Martinovich, and Laura Lopez Ortiz. From the Lemelson Foundation, we thank David Coronado, Carol Dahl, Pam Kahl, Rob Schneider, and Jea Seconi. We thank Mimi Eisen and Jennifer Giambrone of History Associates for sharing a transcript from their interview with Leroy Hood. We thank Joe Hadzima of IPVision for sharing data and information on the prize winner's patent portfolios. Within the RAND Corporation, we thank David Adamson for assistance with communications planning and Melissa Bauman and Nora Spiering for assisting with the layout of the infographics in the appendix. We thank Babitha Balan, Nora Spiering, Jessica Wolpert, and Jessica Arana for shepherding this report through the publication process in a timely fashion. Finally, we thank Jirka Taylor and Petra Moser for their thoughtful reviews of the initial draft of this report, as well as Susan Straus and Lance Tan for overseeing the quality assurance process.

Abbreviations

ADAP	Antibody Detection by Agglutination PCR
AIDS	acquired immune deficiency syndrome
Caltech	California Institute of Technology
CAPTCHA	Completely Automated Public Turing Test to Tell Computers and Humans Apart
CEO	chief executive officer
COVID-19	coronavirus disease 2019
DNA	deoxyribonucleic acid
G-CSF	granulocyte colony-stimulating factor
GMO	genetically modified organism
FDA	U.S. Food and Drug Administration
HIV	human immunodeficiency virus
IL-2	interleukin-2
IPC	International Patent Classification
IPO	initial public offering
ISB	Institute for Systems Biology
LED	light-emitting diode
MIT	Massachusetts Institute of Technology
NAICS	North American Industry Classification System
PCR	polymerase chain reaction
rDNA	recombinant DNA
RNA	ribonucleic acid
SARS-COV-2	severe acute respiratory syndrome coronavirus 2
UCSF	University of California, San Francisco
USPTO	U.S. Patent and Trademark Office
WIPO	World Intellectual Property Organization
WOS	Web of Science

1. Introduction and Context

Inventions, such as new tools, devices, processes, and medicines, have provided significant benefits to society. These inventions help people around the world live longer, healthier, and more-productive lives. They provide new ways to build, move, communicate, heal, learn, and play.[1]

The potentially life-changing benefits of invention make it important to ensure that policies, programs, and initiatives support the development of inventions and their incorporation into the national and international economy. Historical disparities persist—women, people of color, and lower-income individuals are less likely to patent their ideas, excluding their creative solutions from the approaches being brought to market to address today's most pressing challenges (Fechner and Shapanka, 2018). In some cases, systemic barriers faced by certain groups, including racial violence, have limited inventors' potential impacts (Cook, 2014). Encouraging invention is important both for supporting a robust economy and addressing long-standing inequities in the population of inventors; evidence suggests that lack of exposure to role models and support networks reduces the number of girls, minorities, and individuals who grew up in low-income households who obtain patents later in life (Bell et al., 2019).

Inventions provide large benefits to society, but they also have disruptive effects, such as forcing individual workers to change jobs or fields when previously manual work becomes automated (Manyika et al., 2017). The dissemination of novel inventions is sometimes slowed as governments determine how to best regulate them. In some cases, this avoids foreseen or unforeseen damages; in other cases, this delays the provision of large social and economic benefits.[2]

Purpose and Structure of This Report

The purpose of this report is to illustrate the impacts of inventions by providing a careful and impartial assessment of the value provided to society by a select group of inventors—the

---

[1] The U.S. Patent and Trademark Office (USPTO) defines an *invention* as "any art or process (way of doing or making things), machine, manufacture, design, or composition of matter, or any new and useful improvement thereof, or any variety of plant, which is or may be patentable under the patent laws of the United States" (USPTO, 2021). Similarly, the World Intellectual Property Organization (WIPO) refers to an *invention* as "a product or a process that generally provides a new way of doing something, or offers a new technical solution to a problem" (WIPO, undated).

[2] For example, the refusal of the U.S. Food and Drug Administration (FDA) to approve thalidomide for public marketing, despite its approval in other countries, is lauded as having successfully prevented widespread birth defects in the United States in the early 1960s (Kelsey, 1965). On the other hand, as discussed in Chapter 3, Herbert Boyer and Stanley Cohen waited six years for their patent to be approved while the U.S. Supreme Court decided whether living organisms could be patented.

recipients of the Lemelson–Massachusetts Institute of Technology (MIT) Prize. The Lemelson-MIT Prize was one of the most prominent efforts to recognize and reward invention.

This report is not an assessment of the effectiveness or impact of the Lemelson-MIT Prize or similar prizes.[3] Rather, this report uses the inventions of Lemelson-MIT Prize winners as examples to illustrate the impacts that inventions can have on society. We consider the impacts of this group's inventions in aggregate and through three case studies of prize winners. The Lemelson-MIT Prize winners are a nonrandom, carefully selected group of high-achieving inventors; therefore, the impacts that their inventions have on society may be different from the impacts of other inventions, on average.

Measuring the benefits provided by these individuals' inventions can be challenging. While some inventions are sold in marketplaces at known prices and in known quantities, beneficiaries may value the invention beyond its purchase price, and indirect benefits can go unmeasured. Furthermore, it can often take many years, even decades, for the full potential of a new invention to be realized. For well-established inventions, attribution is often difficult: There are often many individuals involved in the process of invention, and even more individuals involved in the production, distribution, and refinement of the resulting products. Furthermore, a single novel invention can, in turn, inspire related inventions, and the benefits of related inventions can be difficult to disentangle from the initial invention. In some cases, novel inventions can spawn entirely new industries through innovation upon initial technological or scientific breakthroughs (National Research Council, 2012).

In this chapter, we provide context for this report, the Lemelson-MIT Prize winners whose inventions we evaluate, and the societal challenges that continue to limit the potential impact of inventions. Chapter 2 illustrates the impact of inventions by discussing the technical, scientific, economic, and social impacts of the Lemelson-MIT Prize winners' inventions in aggregate; additional details on the impacts of every recipient's inventions are available in a separate appendix (RR-A838-2, *Measuring the Value of Invention: The Impact of Lemelson-MIT Prize Winners' Inventions—Appendix*).

Where possible, we use the same metrics to describe each inventor's work. The purpose of these metrics is not to rank the inventors by their relative impact, especially because newer inventions may have yet to realize their full potential under these measures. Instead, the purpose of these metrics is to facilitate an understanding of the potential value of invention more broadly and to reflect upon the impacts of the prize winners' inventions as a group. We consider both inventions made prior to inventors' receipt of the award and inventions made after their receipt of the award.

Chapters 3 through 5 of this report focus on the impacts of the inventions created by winners of the Lemelson-MIT Prize from three particular years: 1996 (Herbert Boyer and Stanley

---

[3] For studies showing that *ex post* technology prizes can increase the future patenting rate of prize winners, see Brunt, Lerner, and Nicholas, 2012; and Moser and Nicholas, 2013.

Cohen), 2003 (Leroy Hood), and 2010 (Carolyn Bertozzi). These case studies allow us to provide more-specific detail on impacts and associated context than can be provided in the appendix. These case studies were selected by the research team, in collaboration with the Lemelson Foundation and Lemelson-MIT Program, based on the following criteria:

- a well-defined portfolio (e.g., not a single product, a too-diffuse collection)
- an innovative and well-recognized contribution (i.e., through citations and other references)
- sufficient time on the market to observe and measure impacts (i.e., selecting established products in favor of more nascent technologies)
- illustrating the diverse variety of paths that inventors can pursue while benefiting society, even within a common sector (biotechnology)

In all cases, we emphasize that we are describing part of a larger story and are focusing on describing the impact of inventions, rather than the impact of the inventors themselves. Many winners of the Lemelson-MIT Prize are prolific inventors and have made valuable contributions beyond those we describe in this report. In many cases, these inventors have made major contributions through channels beyond their own inventions, such as major charitable contributions and mentoring the next generation of inventors. We also emphasize that, in addition to the Lemelson-MIT Prize winners, many inventors are making important and valuable contributions to society. Other studies, like Roberts and Eesley (2009), have examined the impact of similar groups, such as MIT alumni entrepreneurs.

Chapter 6 offers a concluding perspective on the lessons learned from this study. A separate appendix (RR-A838-2, *Measuring the Value of Invention: The Impact of Lemelson-MIT Prize Winners' Inventions—Appendix*) is available at www.rand.org/t/RRA838-1.html.

About the Lemelson-MIT Prize

The winners of the Lemelson-MIT Prize are a selected group of highly accomplished inventors. This section presents information on the variety of disciplines and categories of inventions represented by this group.

The Lemelson-MIT Prize was one of the most prominent efforts to recognize and reward invention. The $500,000 award, which was the largest cash prize for invention in the United States, was funded by The Lemelson Foundation and administered by the School of Engineering at MIT. The Lemelson Foundation was established by Jerome and Dorothy Lemelson in 1992 with the goal of cultivating future generations of inventors to create a better world. Jerome Lemelson was one of the most prolific independent inventors in the United States, amassing more than 600 patents.[4] The Lemelson-MIT Prize was created to recognize highly accomplished and highly promising mid-career inventors who have developed patented products or processes

---

[4] For more information about Jerome and Dorothy Lemelson and the Lemelson Foundation, see Lemelson Foundation, undated.

that have been commercialized and attained or have a high potential for wide adoption and large societal benefits. The prize was awarded annually from 1995 through 2019, with a total of 26 winners (one winner each year, with the exception of Herbert Boyer and Stanley Cohen sharing the 1996 award). The Lemelson-MIT Program discontinued the prize in 2019 to focus on its efforts on inspiring and encouraging young inventors, including growing its commitment to incorporating invention into K–12 and higher education.[5] As part of these efforts to support invention, the Lemelson-MIT Program leadership asked the RAND Corporation to conduct this independent study measuring the impact of the prize winners' inventions on society.

The inventions of Lemelson-MIT Prize winners span a variety of disciplines and sectors, such as health, technology, energy, manufacturing, and others. Table 1.1 summarizes the wide-ranging contributions of these inventors across sectors based on the North American Industry Classification System (NAICS).[6] Nanotechnology is the most heavily represented sector, with eight prize winners, followed by biotechnology (except nanobiotechnology; four prize winners), computer and peripheral equipment manufacturing (three prize winners), medical devices (three prize winners), and semiconductors and electronic devices (three prize winners).

Table 1.1. Lemelson-MIT Prize Winners by Industry

Industry	Number of Award Winners
Nanotechnology	8
Biotechnology (except nanobiotechnology)	4
Computer and peripheral equipment manufacturing	3
Medical devices	3
Semiconductors and electronic devices	3
Computer programming and software publishing	2
Storage battery manufacturing	2
Audio and other communications equipment	1
Data processing and related services	1
Electronic component manufacturing	1
Industrial machinery manufacturing	1
Measuring devices	1

NOTES: Individual award winners can have inventions in more than one industry. The industry descriptions correspond to the most closely affiliated six-digit NAICS industry code or a combination of similar NAICS codes.

---

[5] Additional information about the 26 winners of the $500,000 Lemelson-MIT Prize and the collegiate inventors receiving the Lemelson-MIT Student Prize can be found at Lemelson-MIT, undated-a and undated-c.

[6] According to the U.S. Census Bureau, "[t]he North American Industry Classification System is the standard used by Federal statistical agencies in classifying business establishments for the purpose of collecting, analyzing, and publishing statistical data related to the U.S. business economy" (U.S. Census Bureau, undated).

Nearly half of the Lemelson-MIT Prize winners produced inventions in the field of research and development in the physical, engineering, and life sciences.[7] However, the types of inventions are much broader than those categories suggest. Prize winners' inventions include electronic devices to facilitate reading and communication, sustainable energy storage systems, and computational processes and programs (including open-source technologies). Table 1.2 presents a brief description of seven Lemelson-MIT Prize winners and some of their notable inventions. More examples are available in the separate appendix.

Table 1.2. Sample of Lemelson-MIT Prize Winners and Inventions

Inventor Name	Description of Inventions	Overview of Impacts
Douglas Engelbart	Invented the computer mouse	The computer mouse, hypertext systems, windows, cross-file editing, groupware, and a host of other Engelbart technologies form the basis of interactive, collaborative computing.
Raymond Kurzweil	Invented the first reading machine for the blind	The Kurzweil Reading Machine reads printed and typed documents aloud, improving quality-of-life for vision-impaired people and advancing the field of automated pattern recognition.
Dean Kamen	Invented the iBOT, a battery-powered wheelchair that can climb stairs, and the Segway	Electric-powered personal transportation increased mobility for those unable to walk and allows those in wheelchairs to have eye-level conversations with those not in wheelchairs.
Nick Holonyak	Invented the first practical light-emitting diode (LED)	LEDs are now ubiquitous, used in devices ranging from alarm clocks to traffic lights.
Stephen Quake	Commercialized inventions revolutionizing human health	Technologies to enable accurate deoxyribonucleic acid (DNA) sequencing support broad research, drug discovery, and diagnostic development activities.
Sangeeta Bhatia	Created tiny technologies for medicine	A noninvasive technology using synthetic biomarkers detects disease activity inside the body, and in vitro models of the liver facilitate long-term hepatic metabolism and toxicity studies, replacing the need for human or animal testing.
Cody Friesen	Created renewable energy technologies	Friesen's low-cost, zinc-air rechargeable battery uses oxygen from the atmosphere to provide a scalable, renewable, and reliable source of power with reduced carbon dioxide emissions, and high-efficiency solar technology generates energy to convert water vapor from the air into drinking water.

SOURCE: Lemelson-MIT, undated-b.

---

[7] Industries in this category include NAICS 541713: research and technology in nanotechnology; NAICS 541714: research and technology in biotechnology (except nanobiotechnology); and NAICS 541715: research and development in the physical, engineering, and life sciences (except nanotechnology and biotechnology); U.S. Office of Management and Budget, Economic Classification Policy Committee, 2017.

In addition to—or often, in the process of—providing the world with new tools, devices, processes, and medicines, the winners of the Lemelson-MIT Prize have also made wide-ranging scientific contributions. Table 1.3 depicts prize winners' scientific publications' top ten Web of Science (WOS) categories.[8] The Lemelson-MIT Prize winners, in aggregate, have contributed at least one scientific publication to 147 different WOS categories.

Table 1.3. Lemelson-MIT Prize Winners' Publications by Top Ten Web of Science Categories

Web of Science Category	Number of Publications	Share
Chemistry, multidisciplinary	709	10.5%
Physics, applied	668	9.9%
Multidisciplinary sciences	490	7.3%
Materials science, multidisciplinary	475	7.1%
Chemistry, physical	416	6.2%
Nanoscience and nanotechnology	341	5.1%
Biochemistry and molecular biology	329	4.9%
Physics, condensed matter	254	3.8%
Polymer science	210	3.1%
Cell biology	172	2.6%

Inequity Limits Societal Benefits

Invention does not exist within a vacuum. Before assessing the impacts of inventions, it is important to recognize that the process of translating inventions to impact—as we discuss in Chapter 2—is affected at each step by the environment in which an idea is realized and developed. As discussed in Chapter 2, many factors affecting the market viability and dissemination of an invention are beyond an inventor's control.

In the U.S. context, quantifiable inequities in the representation of inventors of different race, gender, and socioeconomic status are reflected in the U.S. patent record. Evidence suggests that individuals' likelihood of developing a patentable invention and likelihood of being able to bring that invention to intended audiences is affected by broader social conditions and context. For example, the patent record reveals that between 1870 and 1940, rises in racially motivated violence were related to declines in the patent productivity of Black inventors (Cook, 2014). Studies of the role of socioeconomic class in innovation reveal that "Lost Einstein" children from families with an income below the national median fall from the innovation pipeline despite high

---

[8] The WOS Core Collection is a comprehensive citation index for scientific and scholarly research. It is a curated collection of more than 21,000 peer-reviewed scholarly journals published worldwide in more than 250 science, social sciences, and humanities disciplines. It was originally produced by the Institute for Scientific Information and is currently maintained by Clarivate, formerly the Intellectual Property and Science business of Thomson Reuters (Clarivate, undated-c).

math and science scores and are ten times less likely to hold patents than their peers in the top 1 percent of parent income (Bell et al., 2019). Furthermore, as of 2016, the "women inventor rate" had reached only 12 percent (increasing from less than 5 percent in 1976), and growth in that metric has been slow (Toole et al., 2019). Efforts to address these challenges may diversify inventor backgrounds and increase the creativity and impacts of inventions.

2. The Impacts of Lemelson-MIT Prize Winners' Inventions

How Invention Leads to Impacts

Although the process of invention sometimes does begin with serendipitous inspiration—colloquially known as the classic "light bulb" moment—invention (including the invention of the light bulb) is often the result of strategic and organized efforts to develop potential solutions for identified societal needs (Sinfield and Solis, 2016). As highlighted in this report, universities, companies, venture capital firms, and other organizations all play a role in the development of new inventions. Once a new tool, product, or process has been developed, a variety of legal mechanisms govern the assignment and transfer of ownership rights of an invention. The property rights aspects of the invention process—and the debate about the extent to which they stimulate or constrain invention—are largely beyond the scope of this report.

Logic Model

Once something has been invented, there are diverse pathways that can and have been followed to bring the invention from initial creation to a novel product that can be widely adopted to benefit society. However, a generalized characterization of these pathways is helpful for thinking about the estimation of societal benefits.

Figure 2.1 provides a generalized logic model illustration of the path from invention to benefit, with a focus on the application of the invention rather than the development of the invention. We then discuss the different manifestations of this generalized process in greater detail. The factors that lead to the initial development of an invention are important, particularly with respect to the equity issues discussed in Chapter 1, but consideration of such factors are beyond the scope of this report.

Figure 2.1. Illustrative Graphical Representation of a Logic Model

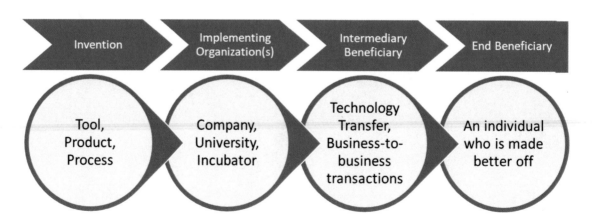

Once an invention has been conceptualized, implementing organizations usually play an important role. If the invention is a physical object intended for mass consumption, copies may need to be produced in large quantities. Physical and virtual infrastructure to support broader use may need to be established or improved to support electronic inventions. In either case, the invention will need to be distributed and marketed or otherwise connected to potential users. Object design and user interaction will also need to be considered, and these may be incorporated into the product development process (Cagan and Vogel, 2002). Although inventors can play a key role in the promotion of their inventions, design, production, and distribution functions are typically addressed by new or existing organizations, which may or may not be led by said inventor.

Often, many organizations are involved in the development of a new marketed product. Although the organization producing and selling the product is often a company, in some cases, a university or incubator may be involved in the initial development of the invention before transferring production rights to another organization. The spread of open-source technology, born from business models developing and distributing open-source software, presents another model in which nonprofits, crowd sourcing, and volunteerism play key roles in helping an invention achieve scale and impact. Many different models may be used to convert an invention into a marketed product; therefore, the phrase "implementing organization" in Figure 2.1 is inclusive of a wide variety of different models.

Implementing organizations help connect inventions to beneficiaries. In some cases, an organization may provide the invention directly to end beneficiaries—i.e., individuals—who use the product to improve their health or overall well-being (a business-to-consumer model), in which case the development of the invention for widespread use flows directly from the second to the final phase depicted in Figure 2.1. In other cases, the invention may be targeted at supporting other organizations and their inventions, such as by making the development of new inventions faster and easier or by increasing the effectiveness of existing inventions (a business-to-business model). In that case, the ultimate benefits to society are indirect; the invention supports activities that provide social and economic benefits. In many cases, these indirect benefits can be quite large because the original invention supports a wide variety of other inventions.

For example, the 2000 Lemelson-MIT Prize winner, Thomas Fogarty, was working as a scrub technician in Good Samaritan Hospital of Cincinnati in the 1940s when he noticed that blood clot surgeries were complicated and often resulted in death. Convinced there must be a better way, he spent many years perfecting his invention, the balloon catheter. Once he perfected the device, Fogarty initially struggled to find an organization that would collaborate with him on their production. "I couldn't get any manufacturer to make the catheters," Fogarty explained. "Companies thought I was some stooge fooling around. I didn't have any credibility." Success only came after Al Starr, the head of the cardiothoracic division at the University of Oregon where Fogarty was in residency, successfully used Fogarty's catheter. Starr convinced Lowell

Edwards, who had successfully collaborated with Starr to manufacture a different cardiovascular medical device, to also manufacture Fogarty's device. Today, Edwards Lifesciences still manufactures Fogarty's balloon catheter, which it sells to medical professionals worldwide. Ultimately, the balloon catheter is estimated to have saved the lives or limbs of as many as 15 million patients and has inspired other improvements in minimally invasive surgeries.[1]

The 2018 Lemelson-MIT Prize winner, Luis von Ahn, began working on a cybersecurity technology in 2000 after Yahoo! Chief Scientist Udi Manber gave a talk to computer science students at Carnegie Mellon University on ten problems that Yahoo! was trying to solve. One problem was that users were signing up for free e-mail accounts offered by Yahoo!, but, at the same time, spammers were writing programs that could rapidly sign up for millions of accounts that could then be used to send spam and other malicious content.

Von Ahn began working with his adviser to develop a verification process to distinguish human input from an automated computer program to thwart spam and abuse. The resulting invention was the Completely Automated Public Turing Test to tell Computers and Humans Apart, or CAPTCHA, a challenge-response test that could be used as a verification process to prevent automated sign-ups. A CAPTCHA test displays fuzzy or distorted letters or numbers and requires users to type out the words or characters shown to solve the test. Most people can read the distorted characters, but automated programs have a harder time distinguishing them. Von Ahn gave the technology to Yahoo! at no cost (Lansat and Feloni, 2018).

Rao and Reiley (2012) estimated that spam advertising cost U.S. firms and consumers approximately $20 billion per year, even after the amount of spam was reduced by CAPTCHAs—and, since then, spamming groups have found ways to solve CAPTCHAs. The second-generation technology developed by von Ahn, reCAPTCHA, blocked automated programs to protect websites from spam and abuse while using human input—users solving millions of CAPTCHAs every day—to help identify text and images.

The reCAPTCHA process is used to digitize books and newspapers, annotate images, and build machine learning datasets. reCAPTCHA can decipher scanned text that could not be read by optical character recognition programs; annotate images, using the computational power of humans to perform a task that computers could not easily do (e.g., character and image recognition); and create difficult-to-collect metadata (e.g., text and labels) for machine learning datasets. Von Ahn used the technology to digitize the archives of the *New York Times* dating back 150 years, deciphering text at a rate of approximately 440 million words per year. In 2009, von Ahn sold reCAPTCHA to Google, which subsequently has used the technology to digitize paper-based texts for Google Books, interpret house numbers and street signs to improve Google Street View, and identify objects on the road to improve artificial intelligence for Google's Waymo autonomous driving systems (Healy and Flores, undated; Kidd, 2019).

---

[1] The quotes and information in this paragraph, as well as the statistic on the number of patients affected, come from Riggins, 2000. Supporting and confirming information came from National Inventor's Hall of Fame, undated.

At the time of writing, more than 5 million websites used reCAPTCHA to prevent spam and abuse (BuiltWith, undated). CAPTCHAs are ubiquitous on most online servers and are solved more than 200 million times per day (Gugliotta, 2011), providing a unique cybersecurity solution and harnessing the computational power of humans for broader applications.

Pathways of Proliferation

Figure 2.2 illustrates the diversity of pathways through which new inventions can be brought to market and, ultimately, to end beneficiaries. We identified more than 180 companies and institutions affiliated with the 26 prize winners and categorized the role of the inventor and the commercial pathways of those organizations.[2] The selected organizations may not represent the entirety of an inventor's commercial activities; they are intended to document instances of major inventions' interactions with the public for characterization and further analysis.

The following three categories describe the role of the inventor in the relevant organization or institution:

1. **founder:** inventions (or patents) resulted in the formation of a new business entity founded (or co-founded) by the inventor
2. **key operating role:** inventions grew or developed business opportunities for an existing organization or institution in which the inventor had a key operating or tenured role
3. **adviser, mentor, or influencer:** patents or concepts developed by the inventor were adopted or inspired new inventions by their peers or advisees that resulted in the formation of an independent business entity (e.g., through a business incubator), or the inventor served as an adviser or on the board of directors of an existing company

We also identify the following pathways for the proliferation of inventions through various commercial entities:

1. inventions that grow a new business entity into a large or established company[3]
2. inventions that grow a new business entity into a small or start-up company (e.g., securing early stage investment)[4]

---

[2] Categories were created through research team review and assessment of public information provided in individual academic or business profiles (e.g., LinkedIn), curricula vitae, company websites, and interviews or profiles of inventors published in newspapers, magazines, or news websites. We developed a matrix of inventor roles and company pathways and had two research team members classify each organization/institution into one block of the matrix. We used additional sources to resolve any discrepancies.

[3] For this category, inventions must have provided a foundation for a public or private company to grow over time and remain an independent entity that exceeds the U.S. Small Business Administration's threshold for a small business. These thresholds vary by industry by employment size or annual revenue. For example, the threshold for biotechnology (NAICS 541714) is 1,000 employees (U.S. Small Business Administration, 2019).

[4] We define commercial entities that do not fall into the large, combined, or inoperative categories as small or start-up companies, although this definition spans a variety of early- to late-stage investment.

3. inventions that generate a new business opportunity resulting in a combined or restructured company (e.g., through a merger or acquisition) to expand market share or increase shareholder value[5]
4. inventions that contribute to forming a new business entity that does not remain financially viable due to bankruptcy or otherwise becomes inoperative[6]
5. other; inventions that were transferred from the inventor to another entity (such as a company, nonprofit, or the public domain) that were not previously commercialized by a new company or business unit.[7]

Figure 2.2 indicates the number of times that a Lemelson-MIT Prize winner has had a specific organizational role for a given pathway. Our intention is not to indicate the superiority of one pathway over another but to provide a sense of the distribution of pathways that inventions take to reach end users and the role that inventors play in that process. Many inventors have played a role in multiple companies to reach new markets and beneficiaries. If the trajectory of a single company included multiple pathways, such as a bankruptcy followed by an acquisition by another company, we only count the first significant event.

The majority of the Lemelson-MIT Prize winners sought to commercialize their inventions as founders of new companies or as advisers or directors of existing companies, but some have successfully disseminated new inventions in other operating roles, through licensing arrangements, or by sharing discoveries, ideas, or technologies.

---

[5] We define commercial entities that have merged with or been acquired by another company or were spun off from a larger existing organization as *combined companies*.

[6] We define commercial entities with a documented bankruptcy or no current operating income as *inoperative*.

[7] This category may undercount transfers that we do not observe because they are not widely documented in the public domain.

Figure 2.2. Pathways of Invention Proliferation

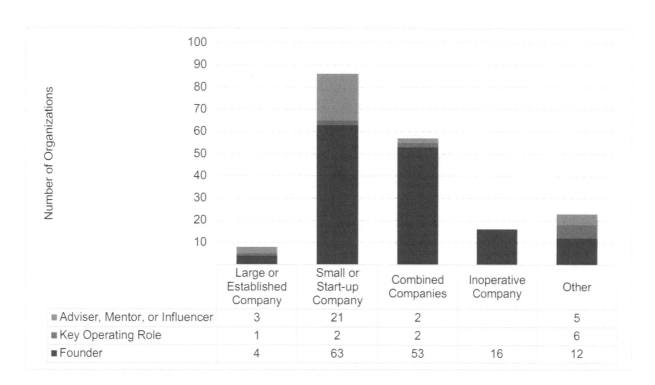

	Large or Established Company	Small or Start-up Company	Combined Companies	Inoperative Company	Other
■ Adviser, Mentor, or Influencer	3	21	2		5
■ Key Operating Role	1	2	2		6
■ Founder	4	63	53	16	12

SOURCE: RAND analysis of data sourced from Yahoo! Finance, Dun & Bradstreet, and other financial news sources.

The most-frequent pathways of invention proliferation among the 26 Lemelson-MIT Prize winners include commercializing an invention as a small or start-up company or having their technology, processes, or operations acquired by a larger company to expand their market share or increase shareholder value. There are a few notable companies—such as Genentech—for which the inventions sustained commercial opportunities that grew into large or established companies.[8] However, the number of individual inventions that gave rise to large companies is relatively small.

This distribution may not reflect the experience of average inventors. Although most prize winners pursued commercialization of their inventions through founding of new (and therefore small) businesses, other inventors may already be working within large businesses (such as the inaugural Lemelson-MIT Prize winner, William Bolander, who worked for General Motors) or pursuing licensing or shared technology arrangements (as Lemelson-MIT Prize winner Douglas Engelbart did for many of his inventions). It is also important to note that entrepreneurship is inherently risky and requires a skillset distinct from the act of invention, as there are many potential pitfalls for new inventors, including the challenge of commercializing and scaling up a

---

[8] In 2009, Genentech was acquired by Hoffman La-Roche, a majority shareholder since 1990. Before its acquisition, Genentech had about 11,000 employees and $13.4 billion in revenue in 2008 (Genentech, 1995–2008).

new idea, technology, or product.[9] Several of the companies founded based on the inventions of Lemelson-MIT Prize winners struggled to financially sustain themselves because of limited demand in small and nascent markets, an existing competitive landscape, or other, unforeseen external or operational challenges associated with bringing a new product to market.[10] We found that many of the Lemelson-MIT Prize winners who have developed successful companies also experienced one or two failures. Those failures were most commonly their first forays into entrepreneurship, and were commonly referred to as difficult learning experiences that provided the experience needed to succeed in subsequent efforts.

One reason that maintaining the long-term viability of an operating organization may be challenging for inventors is because the challenges do not always stem from a problem with the invention itself. One example of these difficulties is Joseph DeSimone's effort to establish a dry-cleaning franchise based on less-toxic, more–environmentally friendly solvents that he had developed.

> [O]ur business team wanted to implement a franchise model . . . everything was going to big chains and brands. And we were going to do that in dry cleaning. So it was a lot of pressure. It was well beyond the technology at this point, right? It's branding, it was operations, it was [a] tough, tough business. (DeSimone, 2013)

A decision by Congress to not pass an anticipated tax credit that would have favored the new technology, along with the dotcom bust of the late 1990s, left the new company short on cash. The company almost secured a new source of funding, but was again disrupted by circumstances unrelated to the invention.

> I'll never forget it. It was Monday morning, September 10th, 2001. I thought we had a deal [for more capital] . . . And, you know, the next morning the world changed: 9/11. And money dried up. We had to disband. (DeSimone, 2013)

Although the company disbanded, DeSimone and his colleagues ensured the underlying technology and the dry cleaning shops using it were able to continue.[11]

> [W]e made provisions to get them the chemistry—to buy their own surfactants and detergents. So we were disbanding, but making sure that they could sustain their business. (DeSimone, 2013)

Looking beyond companies founded by the inventors, our analyses and case studies also found the Lemelson-MIT Prize winners frequently went on to mentor future generations of

---

[9] For a similar study on the impact of entrepreneurship, see Roberts and Eesley, 2009.

[10] For example, Aquion Energy, founded by Lemelson-MIT Prize winner Jay Whitacre to develop his novel saltwater storage battery, struggled in part because the price of lithium-ion batteries, an alternative form of energy storage, dropped far faster than anticipated, reducing the anticipated cost-saving advantage of Whitacre's technology (Temple, 2017).

[11] The organization still exists, but has shifted into medical devices and is no longer associated with DeSimone (DeSimone, 2013).

inventors, serve as advisers to other existing companies, or develop business incubators or venture capital firms to support others' start-up companies.

Measuring Impacts

Now that we have discussed the processes by which we measured how inventions lead to societal benefits, the next task is to discuss how we measure these impacts and benefits themselves. This report reviews social benefits from two perspectives: through this chapter's review of the aggregate impacts of the work of all Lemelson-MIT Prize winners and through the subsequent three chapters' deep-dive case studies on the impacts of the particular inventions of individual prize winners. For aggregate impacts, we examine metrics that reflect scientific, technical, economic, and social impacts. In this section, we discuss those metrics and the methodology we employed to produce them.

We find that the economic and social impacts of prize winners are reflected in the more than 180 organizations affiliated with the 26 Lemelson-MIT Prize winners. A small number of companies grew into large profitable firms, some were acquired by or merged with existing businesses, and the majority remained successful smaller companies or were start-ups attempting to commercialize new ideas or technologies. Many companies did not succeed.

The combined technologies, resources, and capabilities of these organizations achieved significant advances in several areas, including pharmaceutical manufacturing, drug delivery systems, pattern recognition–based computer technology, mobility devices, and energy storage systems. The independent business entities founded by Lemelson-MIT Prize winners (for which financial data are available) in aggregate employ approximately 40,000 people and generate total annual revenues exceeding $54 billion.[12] As of 2020, the market valuations of several publicly traded firms founded by Lemelson-MIT Prize winners ranged from more than $100 million to more than $100 billion.

In some cases, the inventions of the Lemelson-MIT Prize winners not only created new companies but were revolutionary creations that spawned entirely new industries. This broader scientific and technical impact is reflected in the citation of the prize winners' work in scientific publications, new product patents, licensing, or through open-source technologies. As of March 2020, the 26 prize winners had published more than 3,700 articles that had accumulated more than 334,000 citations in the Web of Science citation index. As of January 1, 2020, their 3,871 patents had been cited as prior art by more than 40,000 subsequent patents.

---

[12] Financial data are not available for all companies, including many privately held companies. Several start-up companies reported $0 in annual revenues.

Scientific Impact

Methods

We use counts of journal articles indexed in the WOS database as a measure of scientific output, and, by extension, as a proxy for scientific impact. WOS is a large publication database that contains articles from more than 73 million records from 20,900 journals. Although this measure reflects the many publications that Lemelson-MIT Prize winners have authored and co-authored, we note that our count of publications may be lower than alternative counting methods for several reasons. First, such publications as editorials, conference proceedings, book reviews, and commentaries are omitted from our totals. We count only journal articles because of the relatively stable criteria for publishing within a scientific journal. Although there is some interdisciplinary and interjournal variability in the criteria used for accepting an article for publication, the fundamental criteria for publication is that the study is of high quality and makes a contribution to the scientific discipline in question (Bornmann, 2011).

Second, our means of consolidating the journal articles attributable to a given author might have left certain publications out of each author's sample. To provide a comparable publication count for each inventor, we searched the author field within the WOS database using the author's last name, followed by their first initial. We then identified a publication that we knew to have been written by the target inventor to serve as a confirmed publication. For example, one of Cody Friesen's primary research strands is renewable energy technologies and he has been affiliated with Arizona State University since 2004. Therefore, we counted a 2008 *Journal of Power Sources* article titled "Orthogonal Flow Membraneless Fuel Cell" for which "Friesen, C" has an Arizona State University affiliation as a confirmed publication. Attaining a confirmed publication allowed us to distinguish between the target author and other authors that match the initial author search (e.g., Constance Friesen, a professor of management at a Canadian university). We then used the WOS author profile identification algorithm on the identified author. This algorithm generates a profile for an author based on their name and affiliations. We then filtered out nonarticle publications (e.g., editorials, book reviews, and commentaries); the resulting set of scientific publications constituted the sample of publications for each inventor.

This method, while consistently applied for each author, might have omitted certain publications written by an author if the WOS author profile generation algorithm failed to recognize certain publications as attributable to that author.[13]

---

[13] In the case of Leroy Hood, the WOS author profile generation algorithm significantly undercounts Hood's journal article publications (by approximately 400) because many of Hood's publications are misallocated to other authors with the name "L. Hood." Although we use the WOS data in this chapter for aggregate statistics, we acknowledge that the data undercount Hood's journal article publications in a manner that cannot be easily updated. We use Hood's 2020 curriculum vitae as our point of reference for the Hood-specific statistics presented in Chapter 4. We are not aware of discrepancies for the other inventors, although we emphasize that our numbers might differ from other sources depending on the date of the counting and whether publications other than journal articles were included.

One important limitation to our metric of scientific impact is worth noting. Publication counts fails to account for interarticle heterogeneity in impact. That is, certain articles simply present research results that are more impactful than those contained in other articles. It is worth noting that WOS only includes journals that meet its quality standards.[14] This ensures that certain low-quality (often pay-to-publish) journals that have recently proliferated are not included in our results. Nevertheless, even within WOS-indexed articles, there is bound to be significant interarticle heterogeneity in impact. We address the issue of variability in a given article's impact by also examining prize winners' citations (i.e., the number of instances in which a scientific peer has cited a winner's publication portfolio).

Findings

Table 2.1 summarizes the aggregate scientific impact of the Lemelson-MIT Prize winners. As of March 2020, the prize winners have published more than 3,700 articles that have accumulated more than 334,000 citations in the WOS database. On average, these articles have been cited more than 88 times, which is significantly higher than average (Beaulieu, 2015). These articles have been assigned to more than 8,000 unique WOS Keyword Plus keywords, suggesting substantial breadth in scientific content covered.[15]

Table 2.1. Lemelson-MIT Prize Winners Summary of Scientific Impact, 1955–2020

Scientific articles	3,786
Total citations	334,704
Average citations per publication	88.4
Unique scientific journals	682
Unique Keyword Plus keywords	8,327

SOURCE: RAND analysis of WOS data.

The variety of Lemelson-MIT Prize winners' scientific contribution is also evident when considering the scientific journals in which the prize winners' publications have appeared. In total, the Lemelson-MIT Prize winners have published in 682 distinct journals. Table 2.2 depicts the top 20 journals in which the publications of the Lemelson-MIT Prize winners have been published. Besides spanning a variety of scientific disciplines, these journals constitute many of the highest-ranking scientific journals. For example, the following journals from the prize winners' top 20 journal list ranked in the global top 25 in terms of impact: *Nature* (1), *Advanced Materials* (6), *Nature Communications* (7), *Journal of the American Chemical Society* (11),

---

[14] For more information about WOS's selection process, see Clarivate, undated-a.

[15] Keyword Plus keywords are an algorithm-based method of keyword identification used by WOS to assign keywords to articles based on the titles of the articles that cite the focal article (Clarivate, undated-b).

Angewandte Chemie International Edition (12), *Proceedings of the National Academy of Sciences of the United States of America* (13), *ACS Nano* (22), and *Nano Letters* (25).[16]

Table 2.2. Lemelson-MIT Prize Winners Publications' by Journal

Journal	Number of Publications
Applied Physics Letters	270
Proceedings of the National Academy of Sciences of the United States of America	233
Journal of the American Chemical Society	195
Journal of Applied Physics	115
Macromolecules	106
Advanced Materials	93
Biomaterials	72
Science	66
Angewandte Chemie International Edition	61
Nature Biotechnology	49
Nano Letters	48
ACS Nano	43
Nature Communications	41
PLOS One	40
ACM Transactions on Graphics	39
Langmuir	37
Chemistry of Materials	35
Journal of Controlled Release	34
Journal of Vascular Surgery	34
Nature	34

SOURCE: Clarivate, undated-c.

International co-authorship can be used to measure the extent of international scientific collaboration; it also reflects the increasing global nature of scientific research. We calculated the number of instances in which the focal individual has appeared as an author on a publication with an author that is affiliated with an institution from a country other than the country that

---

[16] Global journal ranking is based on Google Scholars' 2020 calculation of the h5-index, which is calculated as "the largest number h such that at least h articles in that publication [published from 2015 to 2019] were cited at least h times each" (Google Scholar, undated).

hosts the institution with which the focal author is affiliated.[17] Figure 2.3 depicts the geographic distribution of the international collaborations of the Lemelson-MIT Prize winners. Prize winners have co-authored with authors from 62 different countries. Prize winners collaborated most frequently with authors from the United Kingdom (115), China (110), South Korea (106), Germany (77), and Japan (68).

Figure 2.3. International Co-Authorship of Lemelson-MIT Prize Winners

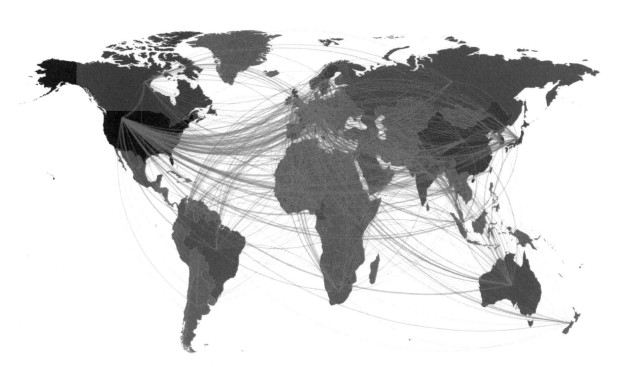

SOURCE: RAND analysis of WOS data.
NOTES: Color shading reflects the number of connections. Darker blues signify countries with more collaborative ties with Lemelson-MIT Prize winners. Countries in gray have no co-authorship ties with Lemelson-MIT prize winners in our articles dataset.

Technological Impact

Methods

We use patents as a measure of technological output, and, by extension, technological impact. To be granted, a patent must demonstrate in the application documents that the underlying invention is novel, nonobvious, and useful.[18] Patent examiners, who have subject-

---

[17] All prize winners are based in the United States.

[18] The USPTO states that "In order for an invention to be patentable it must be new as defined in the patent law, which provides that an invention cannot be patented if:

'(1) the claimed invention was patented, described in a printed publication, or in public use, on sale, or otherwise available to the public before the effective filing date of the claimed invention' or

matter expertise in the field in question, review prior art to verify the application's claims. Thus, patents are a sound proxy for technological invention because they represent the first instance of a nonobvious and useful technology. However, patents, like all measures, suffer from limitations. First, patent-based metrics omit instances of technological invention that are kept as trade secrets. Second, there is significant heterogeneity in patent quality; certain patents represent incremental improvements to the status quo, while others represent radical inventions (Schmid, 2018; Schmid and Fajebe, 2019). Third, patents can be filed strategically to prevent other firms from using an innovation (Blind, Cremers, and Mueller, 2009; Noel and Schankerman, 2013; Thumm, 2004), and even the choices of which citations to cite as prior art may be a strategic decision (Lampe, 2012). Patent attorneys also tend to cite early patents to establish the patentability of an invention (Moser, Ohmstedt, and Rhode, 2018). Nevertheless, the evidence suggests that patent citation counts remain a useful metric, as higher numbers of patent citations are correlated with improved performance of corn hybrids in agricultural settings (Moser, Ohmstedt, and Rhode, 2018).

In this section, we use several different patent-based metrics to describe the technological impacts of the 26 Lemelson-MIT Prize winners. Patents are measured here as the number of patent grants and patent applications on which the individual was listed as inventor as of January 1, 2020. Because of the significant delay associated with the granting of patent applications, using patent applications ensures that recent inventions are included in the individual's patent total.

We note that the patent portfolios and technological contributions of many inventors have grown over time. Many, if not all, of the prize winners have continued to collaborate, mentor, publish new research, and develop new ideas, processes, and technologies after receiving the Lemelson-MIT Prize.

Findings

Counting only patents held by inventors at the date each received the Lemelson-MIT Prize, the 26 inventors collectively owned 1,178 patents. However, each inventor continued to produce inventions after receiving the award. As of January 1, 2020, they held approximately 3,871 patents in total. Table 2.3 reflects several patent-based metrics for each inventor as of January 1, 2020.

'(2) the claimed invention was described in a patent issued [by the U.S.] or in an application for patent published or deemed published [by the U.S.], in which the patent or application, as the case may be, names another inventor and was effectively filed before the effective filing date of the claimed invention.'"

Regarding usefulness, the USPTO states "The patent law specifies that the subject matter must be 'useful.' The term 'useful' in this connection refers to the condition that the subject matter has a useful purpose and also includes operativeness, that is, a machine which will not operate to perform the intended purpose would not be called useful, and therefore would not be granted a patent." (USPTO, 2020)

Table 2.3. Patents, Forward Citations, and Citing Organizations

Year Prize Won	Inventor	Total Patents	Total Forward Citations	Total Citing Organizations
1995	Bolander	14	242	73
1996	Boyer and Cohen	45	690	279
1997	Engelbart	21	270	97
1998	Langer	831	7,487	1,369
1999	Mead	90	2,824	414
2000	Fogarty	215	8,259	795
2001	Kurzweil	117	883	232
2002	Kamen	479	4,560	830
2003	Hood	72	1,377	309
2004	Holonyak	64	613	187
2005	Norris	115	1,432	441
2006	Fergason	133	3,065	624
2007	Swager	117	425	126
2008	DeSimone	206	1,698	403
2009	Mirkin	222	1,267	380
2010	Bertozzi	87	305	90
2011	Rogers	182	2,539	563
2012	Quake	265	2,893	500
2013	Belcher	70	182	90
2014	Bhatia	68	717	168
2015	Whitacre	27	347	81
2016	Raskar	129	1,349	336
2017	Zhang	203	684	139
2018	von Ahn	12	128	49
2019	Friesen	87	37	22
Total		3,871	41,653	5,972

SOURCE: Data from IPVision, Inc. Patent data as of January 1, 2020. Citation data as of July 1, 2020.
NOTE: Forward citations do not include self-citations from an inventor's future works. For forward citations, the sum of the rows exceeds the total because patents may have multiple forward citations (including to more than one inventor on this list); for the total, we avoid double-counting individual patents. For citing organizations, the sum of the rows exceeds the total because organizations may hold one or multiple patents that cite one or more inventors on the list; for the total, we avoid double-counting citing organizations. Citing organizations may include individuals, corporations, and institutions as well as their divisions, subsidiaries, and affiliates. Therefore, some companies or institutions may be counted more than once (for example, if a subsidiary owns a subset of the organization's patents).

Table 2.3 uses forward citations as a measure of the technological impact of an individual's patent portfolio. Patent applicants are required to cite as prior art all patented inventions that were material inputs to the invention under review. The citations from other patents that a patent accumulates over time are referred to as *forward citations*. Patents that have frequently been deemed material inputs to future inventions thus accumulate a large number of forward citations. The fit of forward citations as a proxy for technological impact is supported by the empirical correlation between forward citations and other measures of an invention's import. For example, Albert et al. (1991) find that forward citations correlate strongly to the opinions of knowledgeable peers about the technological import of a given patent. Hall, Jaffe, and Trajtenberg (2005) and Nicholas (2008) find that companies owning highly cited patents have higher stock market valuations. Odasso, Scellato, and Ughetto (2015) find that forward citations correlate with a patent's market value.

Another way to consider the technological reach of an inventor's work is to count the number of distinct organizations that cite their inventions. In Table 2.3, *total citing organizations* measures the number of unique entities that have cited at least one patent within an individual's patent portfolio.[19]

Figure 2.4 provides another way to visualize the scope of the technological impact of the inventions of the Lemelson-MIT Prize winners. During the patent application process, patents are assigned International Patent Classification (IPC) codes by patent examiners to facilitate subsequent applications' search for prior art. IPC codes classify a given invention in terms of its larger technological field. When viewed in the aggregate, IPC codes allow for the characterization of the distribution of group of inventions across technological fields. Figure 2.4 uses IPC codes to group Lemelson-MIT Prize winners' collective patent portfolio into high-level technology sectors.[20] As the figure indicates, the most common technology sector for this group of inventors is instruments, followed by chemistry, electrical engineering, and mechanical engineering.

---

[19] Data are pulled as is from the USPTO. Therefore, *citing organizations* include individuals, corporations, and institutions as well as their divisions, subsidiaries, and affiliates. Therefore, some organizations may be counted more than once.

[20] We rely on the WIPO definitions of *technology sector* and the IPC-WIPO technology sector cross-walk developed by Schmoch, 2008.

Figure 2.4. Distribution of Lemelson-MIT Prize Winners' Patents by Technology Sector

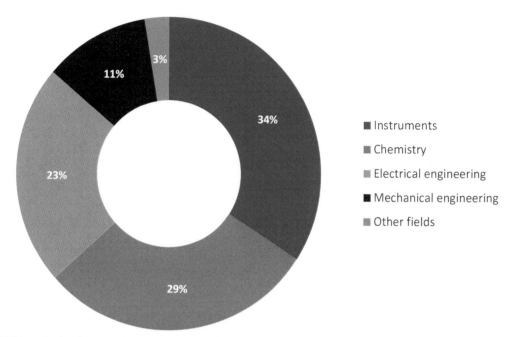

SOURCE: RAND analysis of patent data provided by IPVision, Inc.
NOTE: WIPO technology sector determined by IPC-WIPO sector crosswalk (Schmoch, 2008).

Economic and Social Impacts

We measure aggregate economic and social impacts by looking at metrics involving the companies whose products or services are based on the inventions of the Lemelson-MIT Prize winners. Specifically, we use company revenues as an aggregate measure of the value of goods and services provided. We also estimate the number of jobs provided by these companies. Chapters 3 through 5, as well as the separate appendix, add more detail on the specific social impacts of each invention, such as impacts on health and well-being.

We identified more than 180 organizations affiliated with the 26 Lemelson-MIT Prize winners. The operational status of those organizations is reflected in Figure 2.2. Two companies founded by Lemelson-MIT Prize winners, which are described later in this report, were among the first biotechnology firms to achieve a market capitalization exceeding $100 billion (today there are about a dozen such firms). Several other publicly traded firms founded by Lemelson-MIT Prize winners had market valuations between $100 million and $50 billion as of 2020. Comparable market capitalization data is not available for privately held companies, which make up the majority of companies included in this analysis. As of 2019, 35 other companies founded by prize winners had been acquired by or had merged with other business entities in market transactions valued at approximately $7.5 billion (in 2019 dollars).[21] The combined technologies, resources, and capabilities of these organizations achieved significant advances in several

---

[21] In addition, 22 more of these companies were involved in acquisitions for which our research team was not able to find publicly available financial details.

industries. Of those companies that remained independent entities as of 2019, a small number grew into large profitable companies; the vast majority remained successful smaller companies. Several start-up companies founded by recent prize winners are in early stages of research and development or have yet to fully commercialize new inventions. The commercial value and potential revenue streams of these companies is unknowable at this point in time.

Table 2.4 shows that the independent business entities which report financial data employ, in aggregate, approximately 40,000 people and have total annual revenues exceeding $54 billion.[22]

Table 2.4. Economic Value of Companies Founded by Lemelson-MIT Prize Winners (as of 2019)

Independent companies	
Number of companies	53
Total revenues (in millions)	$54,055
Total employment	40,000
Combined Companies (as of date of first merger or acquisition)	
Number of companies	35
Market value of deals (in millions of 2019 dollars)	$7,460

SOURCES: RAND analysis of data sourced from Yahoo! Finance, Dun & Bradstreet, and other financial news sources.
NOTE: This table only features companies and transactions for which our research team was able to find publicly available financial data; the number of companies listed is smaller than the number referenced in the text and in Figure 2.2 for this reason.

The companies founded by Lemelson-MIT Prize winners have developed hundreds of new processes and products. The combined technologies, resources, and capabilities of these organizations have achieved significant advances in several areas, including pharmaceutical manufacturing, drug delivery systems, pattern recognition–based computer technology, mobility devices, and energy storage systems. Market capitalization and revenues serve as indicators of these types of economic and social impacts, but inventions may have generated positive impacts in other ways. Further, company revenues and market capitalization may not present a complete picture if the value that society places on these inventions exceeds their market value—a concept known as consumer surplus.

Our analysis found that other inventions by Lemelson-MIT Prize winners have provided significant economic value for existing companies in which the inventors played a key operational role, have proliferated through licensing agreements with existing organizations, or have been shared with other business entities, research institutions, or the public (e.g., through open-source technologies) at little or no cost. It is challenging to attribute economic gains to specific inventors or inventions in these contexts—although the inability to do so does not diminish the impacts of the inventions because the benefits may be diffuse and not solely

---

[22] Financial data are not available for all companies, including many privately held companies. Several start-up companies reported $0 in annual revenues.

attributable to a single person or product. Many successful inventors also have gone on to mentor future generations of inventors, serve as advisers to existing companies, and found business incubators or venture capital firms to foster new entrepreneurship.

A detailed description of the benefits associated with every invention of every prize winner is far beyond the scope of any single report. To provide more-detailed examples of investor impacts, the next three chapters discuss the impacts of the inventions of the Lemelson-MIT Prize winners from three particular years: Herbert Boyer and Stanley Cohen (1996 joint winners, Chapter 3), Leroy Hood (2003 winner, Chapter 4), and Carolyn Bertozzi (2010 winner, Chapter 5).

3. Boyer and Cohen: Recombinant DNA Technology

In 1973, Herbert Boyer and Stanley Cohen achieved a breakthrough discovery in microbiology: the invention of a simple, effective method for selecting specific genes from an organism and reproducing this genetic material in practically unlimited quantities. Recombinant DNA (rDNA) technology involves artificially joining or recombining segments of two or more different DNA molecules. These segments then are inserted in bacteria or another type of organism, and these organisms replicate the recombined genetic material (Hughes, 2011).[1]

Before Boyer and Cohen's discovery, it was unknown whether DNA molecules could be successfully cloned in different organisms. Their basic science research technique was widely adopted for numerous practical applications, including the mass production of insulin, growth hormone, and other therapeutic products that could be expressed in bacteria. rDNA technology made it possible to isolate a single gene or DNA segment, sequence it, mutate it in a specific way, reinsert the modified sequence into a living organism, and replicate it. At the time, researchers and journalists noted the enormous potential for genetically engineered pharmaceuticals ("The Gene Transplanters," 1974; "Getting Bacteria to Manufacture Genes," 1974; McElheny, 1974).

Boyer and Cohen's invention gave rise to the field of genetic engineering and laid the foundation for the nascent U.S. biotechnology industry, which would grow significantly in subsequent years. The patented technology was licensed to hundreds of companies, resulting in thousands of new products. Boyer, a founder of Genentech, and Cohen, a scientific adviser for Cetus Corporation, helped commercialize the technology as pioneers among the first biotechnology firms. The two firms collectively employed thousands of scientists, generated billions of dollars in revenues, and would have some of the largest public offerings in U.S. history at the time. In 2009, Genentech would become one of the first biotechnology companies to achieve a market valuation exceeding $100 billion.

By 2020, the entire U.S. biotechnology industry had approximately 2,200 firms, 288,000 employees, and $108 billion in annual revenues (IBISWorld, 2020). Millions of Americans—and substantially more individuals globally—have benefited from access to rDNA products. Recombinant therapeutics have resulted in safer, more-effective, and previously unavailable treatments for a variety of conditions. These conditions include diabetes; growth hormone deficiency; leukemia; colorectal, kidney, and ovarian cancers; hereditary disorders, such as cystic fibrosis; acquired immune deficiency syndrome (AIDS); hepatitis B; hepatitis C; and multiple

---

[1] Much of the background information for this chapter comes from Hughes, 2011, which contains interviews with Boyer and Cohen and others in the biotechnology industry and academia.

sclerosis (Khan et al., 2016). Today, genetically modified organisms (GMOs) in agriculture and medicine are ubiquitous.

Boyer and Cohen's invention benefited from the convergence of several contemporaneous discoveries in molecular biology and fruitful collaborations with other researchers. However, dissemination of the technology faced numerous obstacles, including technical research and ethical challenges, debate about the role of inventors and universities in privatizing and profiting from basic biological research, and legal questions about the patentability of genetically modified organisms. Although the potential for practical applications was far-reaching, the pharmaceutical industry was initially hesitant to invest in genetic engineering and not all commercial partnerships were successful. At the same time, there was fierce competition in the academic community and business world to push the boundaries of the new technology and a race for empirical results and applications. Commercialization would involve multiple challenges, including failed experiments and clinical trials, and there was no initial guarantee of successfully industrializing rDNA technology.

In this chapter, we focus on the scientific, technological, and economic and social impacts of Boyer and Cohen's invention of rDNA technology.

Overview of Inventions

In 1963, Boyer earned his Ph.D. from the University of Pittsburgh and began a postdoctoral fellowship at Yale University in microbiology, focusing on genetic exchange and recombination in bacteria. In 1966, he became an assistant professor in microbiology at the University of California, San Francisco (UCSF). By the late 1960s, evidence was emerging that certain restriction enzymes could snip DNA at specific sequences in the molecule (Danna and Nathans, 1971; Linn and Arber, 1968; Meselson and Yuan, 1968; Smith and Wilcox, 1970). Boyer and others envisioned that restriction enzymes would be useful for cutting, recombining, and characterizing DNA. In 1972, Robert Yoshimori, a graduate student in Boyer's laboratory, isolated the restriction enzyme *EcoRI*, which could cut DNA strands predictably and reproducibly at specific DNA sequences, producing cohesive ends that could bond to complementary DNA strands (Yoshimori, Roulland-Dussoix, and Boyer, 1972). This process represented a leap forward in DNA splicing, reducing a once-complex process to a single step.

In 1960, Cohen graduated from medical school at the University of Pennsylvania. He held several internships, and he completed a postdoctoral fellowship in molecular biology at the Albert Einstein College of Medicine in 1967. During this fellowship, he conducted research on plasmids—tiny rings of DNA in bacterial cells that typically carry antibiotic resistance genes that pass from one bacterium to another. In 1968, Cohen became an assistant professor of medicine at Stanford University, where he continued to work on plasmid isolation and characterization. By mid-1972, Cohen had developed a system for cutting and inserting single molecules of plasmid DNA into bacteria to study plasmid structure and antibiotic resistance. However, this process was inefficient—it produced random length DNA fragments, and the plasmid DNA only rarely entered bacterial cells.

At the same time, other researchers were making scientific advancements in microbiology and biochemistry for manipulating and characterizing DNA molecules. In 1972, Paul Berg developed a biochemical method for joining DNA fragments from different sources at the Stanford Department of Biochemistry. He inserted viral DNA into bacterial DNA, making the first rDNA molecules in a test tube. The approach was technically complex and required numerous enzymes, so most research labs were not equipped to recreate it. In the late 1960s, Walter Gilbert was researching how DNA molecules duplicate themselves at Harvard University; in the early 1970s, he was isolating and sequencing DNA fragments (Gilbert and Dressler, 1968; Gilbert and Maxam, 1973). By 1977, Gilbert had developed a new chemical procedure for sequencing DNA (Maxam and Gilbert, 1977). In other research that would benefit Boyer and Cohen's work, Sharp, Sugden, and Sambrook (1973) developed a method for staining DNA fragments with a fluorescent dye that made the fragments stand out on electrophoresis gels in ultraviolet light—eliminating otherwise lengthy staining and destaining procedures. However, no one else by the early 1970s had created a method for cloning DNA.

In November 1972, Cohen organized a conference on plasmid research in Honolulu, Hawaii; he invited Boyer when he became aware of Boyer's unpublished work on the new restriction enzyme. As Boyer presented on the properties of *EcoRI*, Cohen recognized that this restriction enzyme could be a potential solution to the challenges of inserting plasmid DNA into bacteria for cloning. One evening after the conference, Boyer and Cohen met at a local deli and agreed to a collaboration to construct and clone plasmids containing novel combinations of DNA fragments, acknowledging their shared interest and the benefit of utilizing their combined expertise and resources.

In January 1973, Boyer and Cohen began their first experiment with the help of Robert Helling, who was developing electrophoresis techniques in Boyer's lab, and research assistant Annie Chang, who (among other tasks) transported plasmid samples between Boyer and Cohen's laboratories in her Volkswagen Beetle (Hughes, 2009). In March 1973, Boyer and Helling completed the experiment; they observed a band composed of two types of plasmid DNA fragments in bright fluorescent dye on electrophoresis gels, confirming that they had successfully recombined DNA and cloned it.

The first paper describing how genetically engineered plasmids could be developed by joining together fragments derived from separate DNA molecules and replicated was published in November 1973 (Cohen et al., 1973). Figure 3.1 shows a diagram of Boyer and Cohen's rDNA procedure. Figure 3.2 illustrates the insertion of rDNA into bacteria for cloning.

Figure 3.1. Diagram of Boyer and Cohen's Recombinant Deoxyribonucleic Acid Procedure

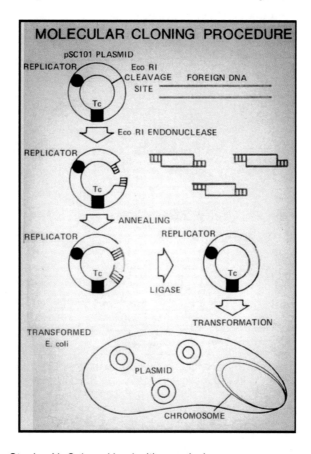

SOURCE: Figure provided by Stanley N. Cohen. Used with permission.

Figure 3.2. Illustration of Insertion of Recombinant Deoxyribonucleic Acid into Bacteria for Replication

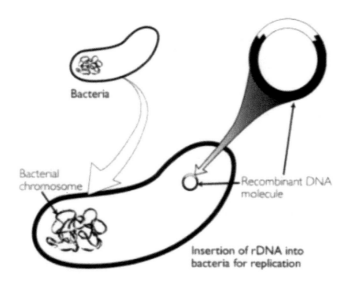

SOURCE: Science History Institute, 2017. Used with permission, courtesy of the Science History Institute.

By summer 1973, Cohen and Chang were working on a second experiment to join and replicate plasmid DNA from two unrelated bacterial species.[2] The recombinant plasmid DNA with antibiotic resistance characteristics of both unrelated organisms was successfully cloned in bacterial cells, suggesting that scientists could replicate foreign DNA in bacteria for various new and useful functions—for example, Cohen conjectured, photosynthesis and antibiotic production (Chang and Cohen, 1974). In July 1973, Boyer and Cohen's team—with the addition of Howard Goodman of the Biochemistry and Biophysics Departments at UCSF and John Morrow, a graduate student of Paul Berg at Stanford—conducted a third experiment to determine if the technology could clone the DNA of a complex organism.[3] The recombinant plasmids were able to clone frog DNA in simple bacterial cells (Morrow et al., 1974). The successful experiment suggested the method could work using the DNA of any species.

Table 3.1 describes the resulting invention from Boyer and Cohen's collaboration and its key beneficiaries.

Table 3.1. Overview of Boyer and Cohen's Invention

Invention	Beneficiaries	Associated Organizations
rDNA technology	rDNA technology gave rise to the field of genetic engineering and laid the foundation for the biotechnology industry, opening the door to gene therapy and the manufacturing of new drugs and existing synthetic biomolecules at scale. Today, thousands of U.S. biotechnology companies use rDNA technology to produce therapeutic products, including insulin, human growth hormone, and the hepatitis B vaccine.	Stanford University and the University of California (technology licensed to hundreds of firms, including Genentech, founded by Boyer)

Stanford University's Office of Technology Licensing, among others, noted the invention's enormous potential for genetic engineering of pharmaceuticals, including insulin and antibiotics. However, there was considerable debate regarding whether scientists could or should patent basic research discoveries. Ultimately, with Cohen and Boyer on board and the University of California as a co-sponsor, Stanford filed a patent application on November 4, 1974.[4] However, the USPTO had ceased reviewing any patent applications involving genetically modified

---

[2] The first experiment used two closely related plasmids that both inhabit *E. coli* bacteria.

[3] At a June 1973 Gordon Conference on Nucleic Acids, Boyer and Morrow had discussed using a sample of purified frog DNA that had been characterized by Morrow's adviser, Donald Brown.

[4] The patent was filed one week ahead of a deadline requiring an inventor to submit an application within a year of the first public disclosure of the invention, which in this case was the publication of Boyer and Cohen's 1973 scientific article (Hughes, 2011).

organisms while a significant legal case on the patentability of living things worked its way to the U.S. Supreme Court. The 1980 outcome of *Diamond v. Chakrabarty*—that "a live, human-made micro-organism is patentable subject matter"—represented a major milestone for the patentability and commercialization of biology and the foundation of biotechnology law (Hughes, 2011).

Shortly after the *Diamond v. Chakrabarty* decision in 1980, patent number 4,237,224, a "process for producing biologically functional molecular chimeras," was the first major patent issued in biotechnology. Within months, more than 70 companies had licensed the patent for their own research and development purposes, agreeing to pay future royalties based on a percentage of downstream drug sales (Hughes, 2011). Stanford would file a second patent four years later, and subsequently a continuation application, for "biologically functional molecular chimeras."[5]

Boyer and Cohen's invention would face other obstacles—primarily concern over the safety of the new technology and academic debate over patenting basic research. At the June 1973 Gordon Conference on Nucleic Acids, Boyer stated that the first experiment with Cohen created a novel combination of antibiotic resistance genes. Conveying unease within the scientific community, a majority of conference attendees voted to send letters to the National Academy of Sciences and the National Institute of Medicine recommending they form a committee to investigate the potential risks of DNA experiments and need for research guidelines. In July 1974, a committee led by biochemist Paul Berg published a letter in *Science* signed by ten prominent scientists—including Boyer and Cohen. The letter called for a temporary moratorium on certain kinds of rDNA research until a conference could be held to consider the risks and develop research guidelines (Hughes, 2011).

In February 1975, the newly formed Recombinant DNA Advisory Committee to Advise the Director of the National Institutes of Health convened 100 molecular biologists at the Asilomar Conference Grounds to consider technical issues of laboratory research safety and develop preliminary draft rDNA research guidelines.[6] By 1976, Congress was considering several bills regulating rDNA research. The political uncertainty delayed many pharmaceutical companies from pursuing genetic engineering research. A decade later, the U.S. Office of Science and Technology Policy published the Coordinated Framework for Regulation of Biotechnology, which detailed government policy for the regulation of new processes and products developed from discoveries using rDNA technology in response to questions about the safety of these

---

[5] A continuation patent allows an owner to pursue additional claims based upon the same description in a prior patent application.

[6] The preliminary draft of new research guidelines was not endorsed by Boyer or Cohen. In June 1976, dozens of pharmaceutical companies expressed concerns to the director of the National Institutes of Health, Donald Fredrickson, about the restrictions on research and development imposed by the new guidelines. The Recombinant DNA Advisory Committee (1982) later issued the revised and less stringent *Guidelines for Research Involving Recombinant DNA Molecules* (Hughes, 2011).

products (Office of Science and Technology Policy, 1986). Meanwhile, the effort by Stanford and the University of California to patent Boyer and Cohen's invention attracted criticism for attempting to privatize and profit from basic biological research (Berg, 1978).

Despite these concerns, Boyer and Cohen's rDNA technology was widely adopted in the scientific community, pharmaceutical manufacturing, and the nascent biotechnology industry. It would radically expand the commercial market for a variety of drugs and therapeutic products, quickly supplanting long-standing methods of producing therapeutic products and treatments, ranging from insulin to growth hormone to vaccines.

Scientific Impact

In 1961, Cohen published his first WOS-indexed article, "Comparison of Autologous, Homologous, and Heterologous Normal Skin Grafts in the Hamster Cheek Pouch" in the *Proceedings of the Society for Experimental Biology and Medicine* (Cohen, 1961). Since 1961, Boyer and Cohen have published an additional 247 WOS-indexed articles, which have appeared in 79 distinct sources. Figure 3.3 depicts Boyer and Cohen's annual publication output from 1960 to 2016, the most recent year in which either author published a WOS-index article.

Figure 3.3. Boyer and Cohen's Web of Science Publications by Year, 1960–2016

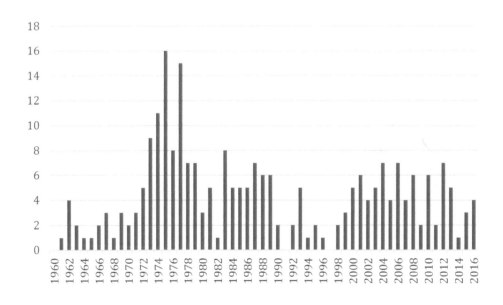

The journals in which Boyer and Cohen's research has appeared most frequently are the *Journal of Bacteriology* (44 articles), *Proceedings of the National Academy of Sciences of the United States of America* (22 articles), and *Molecular Microbiology* (14 articles.) The WOS categories into which Boyer and Cohen's scientific publications are most frequently classified are biochemistry and molecular biology (69 articles), microbiology (49 articles), and multidisciplinary sciences (37 articles). The Keyword Plus terms most frequently assigned to

Boyer and Cohen's scientific publications are "gene" (20 articles), "Escherichia coli" (18 articles), and "protein" (14 articles).

Boyer and Cohen collectively published articles with co-authors from 11 different countries. These collaborations are depicted in Figure 3.4. Boyer and Cohen's top co-authors (frequently researchers in their respective labs at UCSF and Stanford) are Patricia Greene, Mary Betlach, Howard Goodman, John Rosenberg, John Grable, Herbert Heyneker, Francisco Bolivar, Yanan Feng, Joel Hedgpeth, Jianqiang Huang, Kuang-Hung Pan, and Raymond Rodriguez. Boyer generally did not co-author academic publications while at Genentech to avoid the appearance of potential conflicts of interest.

Figure 3.4. Boyer and Cohen's International Collaborations on Scientific Publications

NOTE: Networks graphs are made using the bibliometrix package in R using the following settings: Network layout (star), clustering algorithm (none). Isolates have been removed and the thickness of the edges is weighted to reflect the number of co-authorships between the nodes. The graphs depict all co-authorship dyads, so the Taiwan/United Kingdom tie reflects publications in which at least one Taiwan-based and one United Kingdom–based author were listed on a publication with either Boyer or Cohen.

Boyer and Cohen's scientific publications have been collectively cited approximately 38,000 times by WOS-indexed publications. This amounts to an average of 152 citations per article. Boyer and Cohen's two co-authored scientific articles, "Construction of Biologically Functional Bacterial Plasmids *In Vitro*" and "Replication and Transcription of Eukaryotic DNA in *Escherichia coli*," which were published by the *Proceedings of the National Academy of Sciences of the United States of America* in 1973 and 1974, have been cited 1,172 times and 364 times in WOS publications, respectively (Cohen et al., 1973; Morrow et al., 1974). Boyer's most highly cited article is "Construction and Characterization of New Cloning Vehicle. II. A Multipurpose Cloning System," which was published in in 1977 by *Gene* and has been cited 5,303 times by WOS publications (Bolivar et al., 1977). Cohen's most highly cited article is "Construction and Characterization of Amplifiable Multicopy DNA Cloning Vehicles Derived from the P15A Cryptic Miniplasmid," which was published in 1978 by the *Journal of Bacteriology* and has been cited 3,934 times by WOS publications (Chang and Cohen, 1978). Table 3.2 summarizes the scientific impact of Boyer and Cohen.

Table 3.2. Summary of Boyer and Cohen's Scientific Impact

Total publications	248
Citations	37,610
Sources	79
WOS categories	41

SOURCE: WOS search conducted on March 26, 2020.

Technological Impact

Boyer and Cohen's invention has been identified as one of the most significant patents in the last 180 years. In 2018, a National Bureau of Economic Research study developed an algorithm to measure patent significance using a textual analysis of data from patent documents from the USPTO from 1840 to 2010 (Kelly et al., 2020). The algorithm measured textual similarities between new patents and all existing and subsequent patents to identify patents whose content is novel (distinct from prior patents) but impactful (similar to future patents).[7] The text-based algorithm was designed to construct indexes of technological change at the aggregate, sectoral, and firm-level; however, it also measured the relative impact of individual patents. Out of a list of 250 notable patents awarded since 1840—all of which were within the top 0.5 percent of all patents analyzed in the study—Boyer and Cohen's were identified as the two most influential patents, slightly beating out the telephone (Alexander Graham Bell), one-click buying (Peri Hartman, Jeffrey P. Bezos, Shel Kaphan, and Joel Spiegel), and the airplane (Orville Wright) (Van Dam, 2018).

As of January 1, 2020, Boyer and Cohen collectively held 33 patents and 12 published patent applications. Reflecting both the impact of their technological contributions and the diversity of their practical applications, Boyer and Cohen's patents had 690 forward citations by 279 entities, including individuals, corporations, and institutions and their divisions, subsidiaries, and

---

[7] Alternative methods for measuring the relative impact of inventions rely on patent citations, which were only consistently recorded by the USPTO after 1945, or stock market gains, which are only measurable for patents assigned to publicly traded firms after 1927. A potential limitation of Kelly et al., 2018, is that text-based measures focus on scientific value rather than market value; however, the literature finds these tend to be related (Hall, Jaffe, and Trajtenberg, 2005; Nicolas, 2008; Odasso, Scellato, and Ughetto, 2015).

affiliates. Table 3.3 summarizes the technological impact of the combined patent portfolios of Boyer and Cohen.

Table 3.3. Summary of Boyer and Cohen's Technological Impact

Total patents	45
Forward citations	690
Citing organizations	279

Figure 3.5 provides the distribution of Boyer and Cohen's patents by WIPO technology sector. Boyer and Cohen's collective patent portfolio is largely concentrated in two technology sectors: Approximately 70 percent of Boyer and Cohen's patents are classified under chemistry, and 26 percent are classified under instruments.

Figure 3.5. Distribution of Boyer and Cohen's Patents by Technology Sector

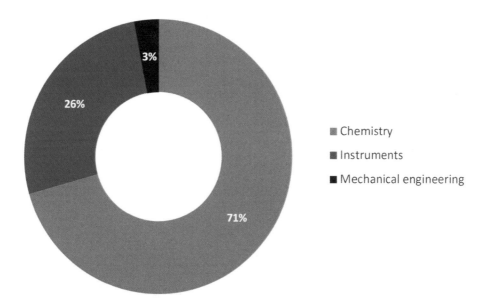

SOURCE: RAND analysis of patent data provided by IPVision, Inc.
NOTE: WIPO technology sector determined by IPC-WIPO sector crosswalk (Schmoch, 2008).

The technological impact of Boyer and Cohen's invention can be seen through the dissemination of rDNA technology and a surge of innovation in the field of genetic engineering by other inventors. This innovation is evident through the forward citations of future patent applications to Boyer and Cohen's invention. The organizations with the most forward citations are Genentech (founded by Boyer), Goldman Sachs,[8] and the United States of America (represented by the Department of Agriculture, the Department of Health and Human Services, and the Department of Veterans' Affairs). The organizations include several biotechnology and pharmaceutical firms, such as Aradigm, Amgen, Chiron Corporation (merged with Cetus Corporation), Eli Lilly and Company, and Illumina. Several research institutions would also make advancements in the fields of genetic engineering and rDNA research by building on Boyer and Cohen's patents, including the University of California, Stanford, Harvard, and MIT. The separate appendix provides a summary of the top citing organizations to Boyer and Cohen's patent portfolio.

Table 3.4 provides examples of new inventions among the top 25 citing organizations (i.e., entities with the highest number of forward citations) to Boyer and Cohen's patent portfolio. These inventions would, in turn, lead to new discoveries and patents, each having dozens to hundreds of forward citations themselves. Although many of these examples are fairly technical and could represent full case studies in and of themselves, the intention of this table is to illustrate the diffuse and reverberating impact of invention.

---

[8] Goldman Sachs Bank USA is the current assignee for 29 patents with forward citations to Boyer and Cohen. It likely took ownership of these patents around the time it financed the acquisition of Gen-Probe Incorporated by Hologic, Inc. in 2012.

Table 3.4. Notable Inventions with Forward Citations from Boyer and Cohen's Patent Portfolio

Citing Organization	Total Patents with Forward Citations	Examples of Notable Inventions	Year
Genentech	31	Recombinant cloning vehicle microbial polypeptide expression	1987
		Methods of producing immunoglobulins, vectors and transformed host cells for use therein	2001
Goldman Sachs Bank USA	29	Nucleic acid probes for detection and/or quantitation of nonviral organisms	1996
United States of America	19	LTR-Vectors	1983
		Recombinant DNA process utilizing a papilloma virus DNA as a vector	1983
Aradigm	16	Creating an aerosolized formulation of insulin	1999
		Use of monomeric insulin as a means for improving the bioavailability of inhaled insulin	1999
University of California, Santa Barbara	16	Codon pair utilization	1992
Amgen	14	DNA sequences encoding erythropoietin	1987
		Production of recombinant erythropoietin	1995
Chiron Corporation	13	Glucoamylase CDNA	1988
Eli Lilly and Company	11	DNA for directing transcription and expression of structural genes	1985
Illumina	10	Massively parallel signature sequencing by ligation of encoded adaptors	2000
Stanford	10	Heat and pH measurement for sequencing of DNA	2011
Harvard	8	Mature protein synthesis	1982
MIT	7	Enhanced production of proteinaceous materials in eukaryotic cells	1987

SOURCE: RAND analysis of data provided by IPVision, Inc.

Economic and Social Impacts

Boyer and Cohen did not collaborate again after 1974. They pursued different paths to commercialize their invention as pioneers among the first biotechnology firms. Both inventors maintained productive labs at their respective universities and would continue to publish, collaborate, and mentor other scientists. Their academic and commercial success would give rise to the field of genetic engineering and lay the foundation for the nascent biotechnology industry.

Cetus Corporation

While remaining at Stanford, in 1975, Cohen joined the science advisory board of Cetus Corporation to advise on applications of rDNA technology. The same year, Cetus Corporation founders Ronald Cape, Peter Farley, and Nobel Prize winner Donald Glaser, a Nobel Prize winner, met with Kleiner & Perkins, a venture capital firm that held a significant stake in the company, to discuss product development (Hughes, 2011). Glaser contemplated recent developments in cloning and the practical applications for genetic engineering, but the other Cetus Corporation co-founders were less interested at the time. Robert Swanson, an associate at Kleiner & Perkins, was captivated with the potential for industrialization of rDNA technology. Ultimately, his failure to persuade Cetus Corporation to found a new genetic engineering division led to Kleiner & Perkins selling their shares in the company and Swanson departing the firm. Swanson would later approach Boyer with a business proposal.

By 1978, Cetus Corporation had finally completed construction of genetic engineering research facilities to conduct their first rDNA experiments. Cetus Corporation, like many other small biotechnology firms, licensed Boyer and Cohen's rDNA technology. The firm developed beta-interferon as a broad-spectrum anti-cancer drug, which was unsuccessful in clinical trials and only later approved to treat symptoms of multiple sclerosis. It also developed interleukin-2 (IL-2), which modifies the immune system for cancer treatment. Cetus applied to patent three new technologies invented by Cohen from 1979 to 1984, including a method for synthesizing DNA sequentially. In March 1981, Cetus Corporation raised $120 million in its initial public offering (IPO), the largest IPO to that date, giving the company a valuation of approximately $500 million (Papadopoulos, 2001). In 1983, Kary Mullis, a DNA chemist at Cetus, invented the polymerase chain reaction (PCR)—a technique for multiplying DNA sequences in vitro. This technique is widely used in DNA research, forensics, and genetic disease diagnostics, and Mullis would win the 1993 Nobel Prize in Chemistry for its development.

In 1990, the FDA declined to approve the company's flagship product, IL-2, for clinical use. That year, Cetus Corporation had $38.9 million in total revenues and $61.5 million in losses (Cetus Corporation, 1990). In 1991, Cetus sold its patent rights to the PCR process to Hoffman-La Roche for approximately $300 million in cash and up to $30 million in royalties and agreed to a merger with Chiron Corporation in a transaction valued at about $660 million in stock (Thayer, 1991). In 1992, Chiron Corporation was able to secure FDA approval for IL-2 for the treatment of advanced-stage kidney cancer. In 1993, beta-interferon was approved by the FDA as the first treatment for multiple sclerosis. Today, several pharmaceutical companies market beta-interferon treatments for multiple sclerosis. Chiron, one of the largest vaccine suppliers in the world, was acquired by Novartis in 2006.

Genentech

Boyer remained at UCSF, joining the Department of Biochemistry and leading a new Division of Genetics; his laboratory focused on cloning animal genes and handling requests for plasmids and restriction enzymes needed for rDNA research (Hughes, 2011). He pursued a research collaboration for an industrial trial of rDNA technology, but that fell through, and he found that pharmaceutical companies had a tepid response to the new technology.

In January 1976, Swanson and Boyer agreed to a partnership and wrote a business plan for industrial uses of rDNA in March 1976, initially focusing on human insulin development. At the time, insulin was created through hormones extracted from cows and pigs, which could cause allergic reactions in some humans. rDNA technology could replicate human insulin, which would pose lower risk of an adverse reaction and mitigate a global insulin shortage. The company's goal would be to license the genetically engineered bacteria to an established pharmaceutical company with the knowledge and resources to develop drugs, conduct clinical trials, and seek the necessary government approvals. On April 7, 1976, Swanson and Boyer signed legal documents incorporating Genentech (Hughes, 2011).[9]

Genentech's early research would be conducted at Boyer's lab and in collaboration with City of Hope National Medical Center in Duarte, California. Boyer's lab at UCSF first collaborated with Arthur Riggs and Keiichi Itakura at City of Hope in 1976 to determine if rDNA technology could be industrialized using synthetic DNA produced by Itakura (Heyneker et al., 1976). Genentech would fund a second collaboration—an experiment to industrialize rDNA technology—in which an artificial copy of a gene able to express somatostatin, a brain hormone that plays a role in regulating growth hormone, was inserted into bacteria (Itakura et al., 1977). Although somatostatin never became a commercial product, the successful experiment demonstrated the feasibility of developing relatively more-complex therapeutic products, such as human insulin.

By 1978, Boyer and Itakura—and an expanding team of postdoctoral researchers and collaborators—had synthesized and purified DNA fragments to insert in plasmids that could express human insulin. Eli Lilly signed a joint-venture agreement with Genentech to license the patent to develop a production process for Humulin, which received regulatory approval from the FDA in 1982. Humulin was the first rDNA product to become commercially available (Hughes, 2011).

With the success of recombinant synthetic insulin, Genentech turned to human growth hormone, which was more complex and had a smaller existing commercial market. In August 1978, Genentech signed a long-term research and development agreement with KabiVitrum, a Swedish pharmaceutical company, to develop bacteria that could express human growth hormone. KabiVitrum was the leading global commercial supplier of human growth hormone, which at the time was a scarce, costly substance extracted from human cadavers. In 1979, David Goeddel and Herbert Heyneker, some of Genentech's first scientists, developed a process for coding human growth hormone by constructing a hybrid gene using a complementary DNA segment. The regulatory approval process was meant to take longer than the approval process for human insulin. However, in 1985, four adults who received cadaver-based growth hormone as children died of Creutzfeldt-Jakob disease. This led the FDA to ban cadaver-derived growth

---

[9] Kleiner & Perkins became the first major investor in Genentech, with Thomas Perkins gaining a seat on the board of directors.

hormone and rapidly approve Genentech's human growth hormone. Genentech would gain exclusive rights to sell recombinant human growth hormone under the brand name Protropin in the United States, while KabiVitrum would sell it internationally, making royalty payments to Genentech (Hughes, 2011).[10]

By the end of 1979, Genentech was pursuing four new projects, three of which were funded by major corporations: interferon, an antiviral protein that moderates the immune system (Hoffmann-La Roche); animal growth hormone (Monsanto); a hepatitis B vaccine (Institut Mérieux); and thymosin, a hormone useful in the treatment of immunodeficiency diseases (Genentech-funded). Genentech was the first biotechnology firm to go public, with an IPO in October 1980. Within an hour of its listing, Genentech's share price rose from $35 to $88, the fastest one-day gain in Wall Street history, before closing at $71.25—giving the company a value of approximately $532 million (Hughes, 2011).

Genentech would go on to develop dozens of new therapeutic products using rDNA technology. In 1990, Hoffmann-La Roche bought a majority stake in Genentech for $2.1 billion, and in 1999, it acquired the remaining portion of the company for about $3.7 billion. However, just months later, Hoffmann-La Roche sold about 34 percent of its shares in two public offerings as Genentech resumed trading on the New York Stock Exchange under a new symbol (DNA). The secondary offering raised $2.9 billion and valued the company at more than $16 billion, which was the largest secondary offering in U.S. history at the time. In 2009, Hoffmann-La Roche acquired the shares of Genentech that it did not already own for $46.8 billion, giving Genentech a valuation exceeding $100 billion (Trefis Team and Great Speculations, 2019). As of 2019, Genentech, a member of the Roche Group, had approximately 13,500 employees, more than 20,000 patents, and more than 40 drugs on the market (Genentech, undated).

Figure 3.6 summarizes Genentech's annual revenues from 1989 through 2008, before the firm was acquired by Hoffman-La Roche.[11] The largest share of revenues came from sales of Genentech's own therapeutic products, which it developed after its early successes in licensing its technology to larger pharmaceutical firms for drug development, clinical trials, and distribution.

---

[10] In 1990, Genentech agreed to a $50 million settlement as part of a criminal lawsuit which alleged that the firm marketed human growth hormone for uses not approved by the FDA.

[11] Around 1990, many U.S. companies first started filing electronic records with the U.S. Securities and Exchange Commission, which does not post earlier historical records online.

Figure 3.6. Genentech Revenues by Year

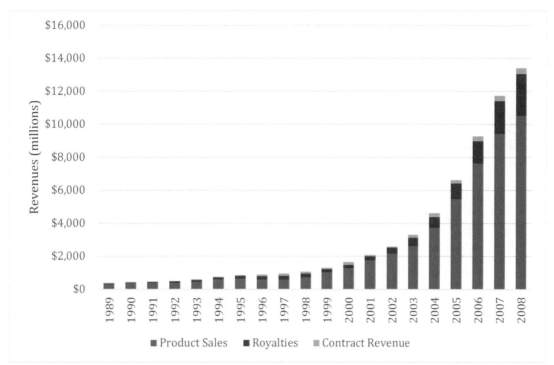

SOURCE: Genentech, 1995–2008.

In 2008, Genentech's top products were Avastin, Rituxan, Herceptin, and Lucentis, which generated sales of about $7.5 billion (about $9.7 billion measured in 2018 dollars). In 2018, the same four products accounted for approximately $21.1 billion in sales for Hoffman-La Roche (Trefis Team and Great Speculations, 2019). Figure 3.7 summarizes Genentech's product sales by year.

Figure 3.7. Genentech Product Sales by Year

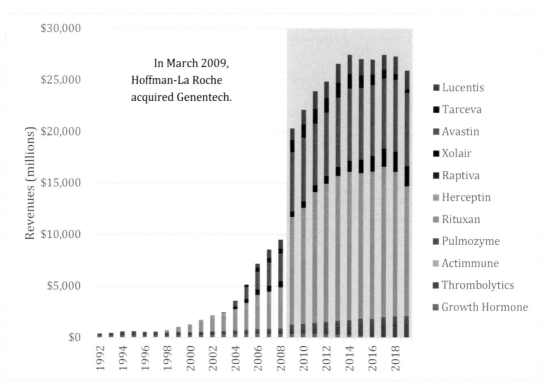

SOURCES: Genentech, 1995–2008; Roche, 2009–2019.
NOTES: Revenues reported in foreign currency were converted to nominal U.S. dollars using an average annual exchange rate. Revenues exclude sales of new drugs introduced after 2008.

Although the sale of pharmaceuticals generated significant private gains for the inventors and their shareholders, they had significant—albeit indirect—impacts for consumers or the end-users of rDNA technology. Millions of Americans—and millions worldwide—have benefited from access to rDNA products (Khan et al., 2016; Steinberg and Raso, 1998). Recombinant therapeutics have resulted in safer, more-effective, and wholly new treatments for a variety of illnesses or conditions. Today, GMOs are ubiquitous in agriculture and medicine (Bessen, 2016).

Table 3.5 describes major therapeutic products developed by Genentech and their impacts. Because of privacy laws and protected proprietary company information, sufficient data are not available to estimate the number of patients that are taking specific rDNA drugs. However, the market size of the biotechnology industry in general, and recombinant drugs specifically, serve as useful proxies for market demand and the perceived value of such treatments. For example, as of 2018, approximately 8.2 million Americans had diabetes; many of them were using one or more formulations of synthetic insulin (Centers for Disease Control and Prevention, undated-a). In North America, the human insulin market size was $10.4 billion in 2018 ("Human Insulin Market Size . . . ," 2020). The National Cancer Institute estimates that 1.8 million new cases of cancer will be diagnosed in the United States and more than 600,000 people will die from the disease in 2020 (National Institutes of Health, National Cancer Institute, 2020). In 2017, the types of cancer associated with highest number of deaths were lung and bronchus (145,849 deaths), colon and rectum (52,547), pancreas (44,011), and female breast (42,000) (Centers for Disease Control and Prevention, undated-b). Estimated national expenditures for cancer care in the United States were $150.8 billion in 2018 (National Institutes of Health, National Cancer Institute, 2020). Modern rDNA drugs and protein therapeutics are used in cancer treatments for millions of Americans, reducing cancer growth, improving remission rates, and reducing cancer deaths.

Table 3.5. Genentech's Major Therapeutic Products and Impacts

Drug or Product	Year Approved	Purpose	Impacts
Humulin	1982	Synthetic human insulin for treating diabetes	Reduced risk of allergic or inflammatory response to animal-based insulin (also alleviated fears over shortage of animal insulin in the 1970s)
Protropin	1985	Treatment for children with growth deficiency stemming from an inability to produce their own growth hormone	• Reduced fatal risk of Creutzfeldt-Jakob syndrome as observed in patients using cadaver-derived human growth hormone • Improved growth outcomes in some children
Activase	1987	Treatment of patients with acute ischemic stroke (sudden stroke)	Reduced incidence of heart failure and reduced risk of mortality
Actimmune	1990	Part of a drug regimen used to treat chronic granulomatous disease	• Reduced frequency and severity of serious infection • Delayed disease progression in patients with malignant osteopetrosis
Pulmozyme	1993	Treatment to manage symptoms of cystic fibrosis	• Improved lung function • Reduced risk of respiratory tract infection
Rituxin	1997	Treatment of multiple diseases, including non-Hodgkin's lymphoma, chronic lymphocytic leukemia, and rheumatoid arthritis	• Prolonged remission in patients with non-Hodgkin's lymphoma • Improved remission rate for acute lymphoblastic leukemia • Reduced signs and symptoms of moderate to severe rheumatoid arthritis
Herceptin	1998	Treatment for early and metastatic breast cancer and stomach cancer	Reduced cancer growth
Xolair	2003	Treatment for allergic asthma and chronic hives with no known cause	• Decreased allergic response • Reduced risk of asthma attacks
Avastin	2004	Treatment for metastatic colon, rectal, and other cancers	• Inhibited growth of new blood vessels to reduce cancer growth • Prevented or slowed vision loss associated with age-related macular degeneration
Lucentis	2006	Treatment for age-related wet macular degeneration	Prevented or slowed vision loss associated with age-related macular degeneration

NOTE: All drug or hormone treatments pose the risk of serious side effects that must be weighed against the potential benefits of treatment.

Stanford University and the University of California

While Boyer and Cohen followed different paths to commercialization, Stanford University and the University of California retained the patent rights to Boyer and Cohen's rDNA technology. Boyer and Cohen's first patent was issued in October 1980. By December 1980, 72 companies had licensed the patent for an annual fee of $10,000 plus downstream royalties based on a percentage of final product sales (Hughes, 2011). Over the lifetime of the patents, which jointly expired in December 1997, the technology was licensed to 468 companies, resulting in more than $35 billion in sales for an estimated 2,442 new products (Feldman, Colaianni, and Liu, 2007). The academic institutions received approximately $255 million in licensing revenues, much of which was subsequently invested in research and research infrastructure. Approximately 90 percent of the revenue came from royalty income on drug sales. In the 1990s, 400 new rDNA products, on average, were introduced to market every year, and sales grew exponentially.

Table 3.6 shows the top licensing firms of Boyer and Cohen's patents. The largest share of royalties came from Amgen ($55 million), Eli Lilly and Company ($37 million), and Genentech ($35 million). In addition to the standard license agreements, alternative licensing agreements were made available to small firms (with less than 125 employees), small distributors or resellers, and start-ups. As a result, the licensed rDNA technology was used by hundreds of small biotechnology firms, some of which grew into large and successful firms.

Table 3.6. Top Licensees of Boyer and Cohen's Patents

Company	Royalty Payments (in millions)	Top Products
Amgen	$54.8	Epogen, Procrit, Neupogen
Eli Lilly and Company	$36.7	Humulin (with Genentech)
Genentech	$34.7	Humulin, Protropin, Activase, Pulmozyme, Actimmune
Schering-Plough	$18.0	Intron A
Johnson & Johnson	$13.4	Procrit
Merck	$10.1	Recombivax HB
Abbott	$9.8	Various in vitro human immunodeficiency virus (HIV) diagnostics
Novo-Nordisk	$8.7	Novolin
Genetics Institute	$5.9	Recombinate
Chiron Corporation (merged with Cetus Corporation)	$5.1	Proleukin, Betaseron

SOURCE: Feldman, Colaianni, and Liu, 2007.

Boyer and Cohen's patents expired at the end of 1997, but rDNA technology has proliferated and been widely adopted to develop new drugs and therapies. The U.S. biotechnology industry has also expanded significantly. As of 2016, the U.S. biotechnology industry had more than 440 publicly traded companies with 135,000 employees and a market capitalization of about $700 billion (EY, 2017). IBISWorld estimates that in 2020, the U.S. biotechnology industry had approximately 2,200 total firms, 288,000 employees, and $108 billion in annual revenues (IBISWorld, 2020).

These economic impacts include the contributions of hundreds of scientists and tens of thousands of others who would collaborate with the inventors, manage and grow their companies, and license and sell their products. The growth of the biotechnology industry that was founded on the potential of rDNA technology would grow in new and unforeseen ways through the contributions of numerous other inventors (including two others profiled in this report). Although these industrywide impacts cannot be solely attributed to individual inventors, they reflect the enormous and prolonged impact of invention. They might not have happened nearly as quickly—or at all—without the determination of the inventors and the risks they took to develop and commercialize their invention.

Conclusion

Boyer and Cohen's rDNA technology accelerated the mass production of synthetic biomolecules and the manufacturing of new drugs at scale, giving rise to the field of genetic engineering and the biotechnology industry. This technology radically expanded the commercial market for a variety of drugs and therapeutic products. It quickly supplanted longstanding methods of producing therapeutic products, such as insulin, growth hormone, and vaccines. Over the lifetime of Boyer and Cohen's patents, which jointly expired in December 1997, the technology was licensed to hundreds of companies, resulting in thousands of new products. Today, it is used by many more.

Boyer and Cohen were pioneers among the first biotechnology firms. The two firms most closely affiliated with the inventors collectively employed thousands of scientists and generated billions of dollars in revenues. In 2009, Genentech would become one of the first biotechnology companies to achieve a market valuation exceeding $100 billion. In 2020, there were approximately 2,200 biotechnology firms with 288,000 employees and $108 billion in annual revenues.

One of Boyer and Cohen's main impacts—which also brought significant controversy—was on the debate over patenting basic research and commercializing inventions outside academia. Boyer, as a founder of Genentech, would face many challenges before successful clinical trials brought new products to market and established the firm's reputation and future commercial success. Cohen, as a scientific adviser at Cetus Corporation, would patent several new technologies and see the company, in 1981, achieve the largest IPO to date—before Cetus

Corporation's flagship product failed to gain regulatory approval for clinical use, resulting in the company selling some of its prized technology and merging with another company. The experience of both inventors—including both the risks and rewards inherent in the commercialization of invention—would lead to venture capital firms seeking out and finding promising academic researchers and technologies for investment and several prominent scientists taking steps to found and join new biotechnology companies in the late 1970s and beyond.

Millions of Americans—and substantially more people worldwide—have benefited from access to rDNA products. Recombinant therapeutics have resulted in safer, more effective, and previously unavailable treatments for a variety of illnesses or conditions. Today, GMOs are ubiquitous in agriculture and medicine. Some of the major economic and social impacts have included new treatments for diabetes, growth hormone deficiency, leukemia, AIDS, hepatitis B, and multiple sclerosis—all of which have reduced the risk of premature death and provided tangible quality of life improvements for many.

4. Hood: The DNA Sequencer and Modern Scientific Instruments

Leroy Hood laid the technological foundation for genomics (the study of genes) and proteomics (the study of proteins) through the invention of some of the core instruments of modern molecular biology. These instruments allowed rapid automated protein and DNA sequencing and led to significant advancements in mapping and sequencing human genes and the discovery of new proteins, which in turn led to improved medical treatments. The scientific instruments invented by Hood and others have allowed researchers to advance the frontier of scientific and technological understanding by automating experimental processes that once were conducted by hand, substantially accelerating the process of discovery. Hood summarizes the catalyzing role of technology in scientific discovery, stating that "with new technologies, biologists have the chance to open up new horizons for exploration in biology" (Hood, 2002).

Hood's tools have affected fields as diverse as biochemistry, molecular biology, cell biology, oncology, biophysics, immunology, nanoscience, neuroscience, zoology, agronomy, parasitology, toxicology, polymer science, and mathematical computational biology. Hood's instruments have been used, among other things, to sequence the prion protein; synthesize oligonucleotides; synthesize the 99 residue HIV-1 protease, contributing to the development of the protease inhibitor to treat HIV/AIDS; and enable the mapping and sequencing of the entire human genome. The Human Genome Project, for which Hood was an early advocate and key contributor, was the largest international collaboration ever undertaken in the field of biology, involving thousands of scientists and billions of dollars of investment by the U.S. government and others.

During his career, Hood was a co-founder, investor, or adviser to more than a dozen companies (including Applied Biosystems and Amgen) that collectively provide more than 40,000 jobs and generate more than $30 billion in annual revenues. Several of these companies combined, through mergers or acquisitions, with other organizations to achieve significant advances in genetic research, pharmaceutical manufacturing, and medicine. Hood later co-founded the Institute for Systems Biology, a nonprofit biomedical research organization, which continues to advance research in fields of study that Hood pioneered, such as systems biology and systems medicine. The impacts and contributions of the institute are still being realized today.

This chapter largely focuses on the contributions of Hood's automated instruments that had significant scientific, technologic, economic, and social impacts.

Overview of Inventions

In 1960, Hood completed a B.S. in biology at California Institute of Technology (Caltech) and enrolled at the Johns Hopkins School of Medicine, where he was granted an M.D. in 1964. Hood then enrolled in the biochemistry doctoral program at Caltech. He defended his doctoral thesis (*Immunoglobulins: Structure, Genetics, and Evolution*) and was awarded a Ph.D. in biochemistry from Caltech in 1968; the defense committee was chaired by molecular biologist William Dreyer, (Hood, 1968). Hood's doctoral adviser underscored the importance of instruments in exploring the scientific frontier, teaching Hood that "if you really want to transform a biological discipline, invent a new technology that permits you to explore new dimensions of data space" (Hood, 2008).

Over the course of his career, Hood conducted research and developed ideas for invention at several institutions. Following the completion of his doctorate, Hood took a position at the National Institutes of Health, a position he held for three years (Hood, 2002). Next, he accepted a faculty position at Caltech, a venue that offered what Hood viewed at the time as "an ideal environment for both biology and technology development" (Hood, 2002). From 1970 to 1992, Hood served on the biology faculty at Caltech, where he worked on four of his most widely recognized and frequently cited inventions: the gas-liquid phase protein/peptide sequencer, the DNA synthesizer, the protein synthesizer, and the DNA sequencer.[1] In 1989, Hood stepped down as chairman of the Division of Biology to become the director of the Center for Molecular Biotechnology at Caltech, a new science and technology center funded by the National Science Foundation. In 1992, he left Caltech to found the first cross-disciplinary biology department and was appointed chair of the Department of Molecular Biotechnology within the School of Medicine at the University of Washington. During the next two years, he transferred the Center for Molecular Biotechnology to Seattle. In 1999, Hood left the University of Washington to found the Institute for Systems Biology (ISB), a nonprofit collaborative and cross-disciplinary biomedical research organization.

Hood's early discoveries in molecular immunology led to a better understanding of the structure and diversity of antibody genes, but his most-notable impacts are largely in the development of scientific instruments, which revolutionized the emerging fields of genomics and proteomics and enabled the Human Genome Project. Table 4.1 summarizes some of the key inventions of Hood and others as well as beneficiaries and associated organizations.[2] Applied Biosystems would commercialize all four inventions.

---

[1] These four instruments do not constitute the full extent of Hood's contribution to scientific instrumentation or methodologies. For example, Hood later invented the ink-jet oligonucleotide arrays synthesizer.

[2] Discretizing the innovative contributions of a researcher as prolific as Hood is inexact, here our assessment is consistent with that of Hood himself. In a 2002 article providing a personal account of his contribution to molecular technology, Hood focuses on the inventions described here (Hood, 2002).

Table 4.1. Overview of Hood's Inventions

Invention	Beneficiaries	Associated Organization
Gas-liquid phase protein/peptide sequencer	The gas-liquid phase protein and peptide sequencer (protein sequencer) allowed researchers to determine the chemical makeup of key proteins and peptides using a small amount of sample.	Caltech and Applied Biosystems
DNA synthesizer	The DNA synthesizer allowed the creation of important reagents for molecular biology, such as DNA probes for DNA mapping and sequencing.	Caltech and Applied Biosystems
Protein synthesizer	The protein synthesizer allowed researchers to produce large volumes of proteins, which, in turn, enabled large-scale experimentation to accelerate scientific discovery at a significantly reduced cost.	Caltech and Applied Biosystems
DNA sequencer	The DNA sequencer was one of the key technologies that enabled the completion of the Human Genome Project, which completed the full mapping and sequencing of the entire human genome in 2003.	Caltech and Applied Biosystems

In addition to the other inventions discussed throughout this chapter, researchers at Hood's laboratory at Caltech were responsible for developing at least five distinct strategies for genomic analysis, including sequence-tagged sites for physical mapping (Olson et al., 1989; Hood, 2002).

Gas-Liquid Phase Protein/Peptide Sequencer

The first gas-liquid phase protein and peptide sequencer (protein sequencer) was developed by Hood, Rodney Hewick, Mike Hunkapiller, and William Dreyer at Hood's Caltech laboratory in the early 1980s. The inventors presented the technical specifications of the instrument in a *Journal of Biological Chemistry* article, which has been cited more than 2,200 times (Hewick et al., 1981).[3] The protein sequencer allows researchers to determine the chemical makeup of key proteins and peptides. The instrument can provide useful protein and peptide sequence data using a small amount of sample—it was 100 times more sensitive than its predecessors (Hood, 2002). This increased sensitivity allowed certain proteins that are only available in low quantities to be sequenced. Hood summarizes the effect of the protein sequencer on the field, stating that "the highly sensitive protein sequencer opened up a multiplicity of new areas in biology through the sequence analyses of heretofore inaccessible proteins and the cloning of their corresponding genes" (Hood, 2002).

The protein sequencer has made scientific contributions in fields as diverse as biophysics, cell biology, immunology, food science, zoology, neuroscience, agronomy, parasitology, toxicology, and polymer science.[4] Hood and his colleagues used the instrument to sequence platelet-derived growth factor (a human blood hormone), which showed that the N-terminal

---

[3] Based on a search of Google Scholar Citations conducted on October 1, 2020.

[4] Based on a citation analysis of Hewick et al., 1981.

sequence of the hormone was highly similar to that of the v-cis avian oncogene. This finding led to the hypothesis that oncogenes are "genes of human growth and development subject to control by a cancer virus" (Hood, 2008). This was the first time that part of a protein sequence was queried against a database of known protein strings, and it marked an important early step in the field of bioinformatics. The automated instrument was used to sequence of the interferon-alpha and interferon-beta proteins and the torpedo acetylcholine receptor, which led to treatments for cancer and multiple sclerosis (Knight et al., 1980; Zoon et al., 1980; Raftery et al., 1980). The protein sequencer was used by Stanley Prusiner, Hood, and others to sequence the prion protein. Prusiner was awarded the 1997 Nobel Prize in Physiology or Medicine for his discovery of the prion, a type of protein that can trigger normal proteins in the brain to fold abnormally, contributing to fatal neurodegenerative disease.

The Applied Biosystems protein sequencers, based on the initial prototype developed at Hood's Caltech laboratory, were also the basis for important scientific findings. The Applied Biosystems Model 470A alone is associated with more than 2,000 scientific publications.[5] For example, the Model 470A was used to sequence the Jack Bean Urease (Mamiya et al., 1985), which was the first enzyme ever to be crystallized (Sumner, 1926). It was also used to sequence locust neuroparsins (Girardie at el., 1989) and luminescent protein aequorin from a jellyfish (Inouye et al., 1985), and it was used to sequence and clone sarcoplasmic reticulum—an intracellular membrane that regulates calcium ions in muscle fibers—in a rabbit (Fliegel et al., 1989). Derynck et al. (1985) used the Model 470A protein sequencer to report the amino acid sequence for transforming growth factor-β (TGF-β) in humans. TGF-β is thought to play a role in wound healing and has been shown to affect cell proliferation.

DNA Synthesizer

Hood developed a prototype DNA synthesizer in collaboration with Susan Horvath, Michael Hunkapiller, and others (Horvath et al., 1987). By automating the phosphonamidite chemistry for DNA synthesis developed by Marvin Caruthers, the DNA synthesizer reduced the time and cost necessary to conduct critical portions of the discovery process (Hood, 2008). The DNA synthesizer has been used to produce important reagents for molecular biology, including DNA probes for DNA mapping and sequencing (Hood, 2002). One of the most important scientific contributions of the DNA synthesizer was facilitating the synthesis of oligonucleotides. The DNA synthesizer was used to assemble oligonucleotides in an important study on genetic disease detection methods (Barany, 1991). The DNA synthesizer played an important role in Kary Mullis's discovery of the polymerase chain reaction DNA amplification procedure, which enables the production of millions of DNA copies from a very small DNA sample. Mullis was awarded the Nobel Prize in Chemistry for this discovery in 1993 (Hood, 2008).

---

[5] Based on a Google Scholar search of "Applied Biosystems Model 470A" conducted on October 22, 2020.

The Applied Biosystems model 380A DNA synthesizer has been used in thousands of scientific studies.[6] For example, it was used in two highly cited papers describing a groundbreaking gene splicing technique known as gene splicing by overlap extension, a variant of polymerase chain reaction (Horton et al., 1989; Horton et al., 1990), and a highly impactful study describing a novel procedure for prenatal diagnosis of genetic diseases (Kogan, Doherty, and Gitschier, 1987). The Model 380A DNA synthesizer was also used in an early study demonstrating the ability to determine the sex of very early human embryos during the in-vitro fertilization process (Handyside et al., 1989).

Protein Synthesizer

Hood, in collaboration with Stephen Kent and others, developed the first automated protein synthesizer capable of building long peptides using amino acids. The protein synthesizer allowed researchers to produce large volumes of proteins, which, in turn, enabled large-scale experimentation to accelerate scientific discovery at a significantly reduced cost. Kent, a researcher in Hood's lab, collaborated with the pharmaceutical company Merck to use the protein synthesizer to synthesize the 99 residue protein from the HIV-1 protease, which allowed for the development of the highly effective protease inhibitor treatment for HIV/AIDS. Hood, Kent, and others used the technology to synthesize a variety of other proteins, such as interleukin 3 (Clark-Lewis, Hood, and Kent, 1988) and zinc fingers (Parraga et al., 1988).

The first commercialized model, Applied Biosystems's 430A peptide synthesizer, is referenced in more than 2,000 scientific publications.[7] For example, it was used to synthesize four monellin analogues (a sweet-tasting protein) (Kohmura, Nio, and Ariyoshi, 1992), porcine cardiodilatin-88 (Nokihara, 1988), and many other peptides (Flynn et al., 1991; Schnölzer et al., 1992; Montague et al., 1994; Talanian et al., 1992).

DNA Sequencer

In the mid-1980s, Hood, in collaboration with Henry Huang, Lloyd Smith, Mike Hunkapiller, and Tim Hunkapiller, developed the automated DNA sequencer, which would have significant impacts in many fields (Hood, 2002). The DNA sequencer allowed scientists to input a sample of DNA and rapidly determine the order of the four bases. The DNA sequencer also allowed rapid genotyping, a related process that focuses on determining the precise genetic variants that an individual possesses. Of the instruments invented at Caltech, Hood describes the DNA sequencer as the most difficult to develop (Hood, 2008). A *Nature* article describing the instrument, which allowed rapid automated sequencing of DNA, has been cited more than 2,300 times (Smith et al., 1986).

---

[6] A Google Scholar search of "Applied Biosystems model 380A" or "Applied Biosystems 380A" conducted on October 26, 2020 yielded more than 2,300 publications.

[7] Based on a Google Scholar search of "430A peptide synthesize" conducted on October 22, 2020.

The techniques for sequencing DNA developed independently by Walter Gilbert and Frederick Sanger in the late 1970s, for which they would share the Nobel Prize in Chemistry in 1980 with Paul Berg, were significant advances in molecular biology, but were costly and labor intensive (Genome News Network, undated). Hood and his colleagues improved upon the existing Sanger method by automating the process, modifying both the chemistry and data-gathering processes. Hood's Caltech lab replaced the use of radioactive labels, which were unstable and posed a health risk to researchers, with fluorescent dyes of different colors—one for each of the four DNA bases. This eliminated the need to run several reactions using separate gels and simplified the data-gathering process. Hood also integrated laser and computer technology to transmit information directly to a computer, eliminating the labor-intensive process of gathering sequence data by hand.

The DNA sequencer was one of the key technologies that enabled the completion of the Human Genome Project, which successfully completed the mapping and sequencing of the entire human genome in 2003. The DNA sequencer also enabled the full sequencing of important animal models, such as the mouse genome, which was completed in 2002 (National Institutes of Health and National Human Genome Research Institute, 2002). The sequencing process was instrumental in early work on DNA typing using human hair, which has been used in the field of forensics (Higuchi et al., 1988). Hood states that the DNA sequencer

> also made possible the current analyses of thousands of genome sequences from microbes, plants, animals, and even multiple humans; these, in turn, have transformed and are transforming many different fields of biology and medicine. (Hood, 2008)

The Model 370A DNA sequencer was commercialized in 1986 by Applied Biosystems. It would have significant impacts in fields as diverse as biochemistry, molecular biology, cell biology, oncology, biophysics, immunology, nanoscience, plant sciences, veterinary sciences, mathematical computational biology, and pediatrics.[8]

Scientific Impact

In 1965, Hood, while a graduate student at Caltech, published his first WOS-indexed article, "Evidence for Amino Acid Sequence Differences Among Proteins Resembling the L-Chain Subunits of Immunoglobulins," in the *Journal of Molecular Biology* (Bennett et al., 1965). Over the course of his career, Hood has published more than 900 peer-reviewed articles.[9] Figure 4.1

---

[8] Analysis based on WOS citation analysis of Smith et al., 1986, conducted on October 23, 2020.

[9] The WOS author profile for Leroy Hood was found to be incomplete, omitting a relatively large number of publications. Therefore, we rely on the "peer reviewed articles" section of Hood's 2020 curriculum vitae for this analysis. The exceptions are the network graph (Figure 4.2) and the aggregate information presented in Chapter 2, for which we rely on Hood's WOS author profile.

depicts Hood's annual publication output from 1965 to 2019 (the most recent year for which complete data are available).

Figure 4.1. Hood's Publications by Year, 1965–2019

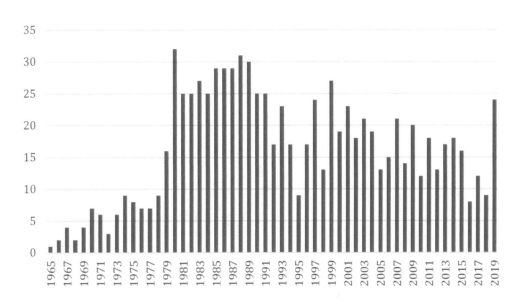

Hood's research has appeared frequently in high-ranking journals, including *Proceedings of the National Academy of Sciences of the United States of America* (more than 100 articles), *Nature* (more than 70 articles), and *Science* (more than 50 articles). Hood's scientific publications span a variety of disciplines, including multidisciplinary sciences, genetics and heredity, biochemical research methods, biotechnology and applied microbiology, and cell biology. One of the top author-provided keywords for Hood's journal articles is "systems biology."

Hood's most-cited publication is the International Human Genome Sequencing Consortium's "Initial Sequencing and Analysis of the Human Genome" (International Human Genome Sequencing Consortium, 2001). Hood's next-most-cited journal article is "A Distinct Lineage of CD4 T Cells Regulates Tissue Inflammation by Producing Interleukin 17," which was published in *Nature Immunology* (Park et al., 2005) and has been cited 2,803 times by WOS publications.

Hood's international collaborative network is extensive. He has co-authored with scholars from at least 35 countries. Figure 4.2 depicts these international research collaborations.

Figure 4.2. Hood's International Collaboration on Scientific Publications

NOTE: Because this figure is based on WOS data, it reflects only a subset of Hood's scientific publications. The networks graph was made using the bibliometrix package in R and the following settings: network layout (circle), clustering algorithm (none). Isolates have been removed and the thickness of the edges is weighted to reflect the number of co-authorships between the nodes. The graphs depict all co-authorship dyads: Ties between non-U.S. countries reflect publications in which at least one co-author from each country was also listed on a publication with Hood. The network graph depends on data from Hood's WOS-generated author profile. As described earlier, we found Hood's author profile to undercount his scientific articles. Because the author profile omits many of Hood's publications, this network graph likely underrepresents the extent of Hood's international collaborations.

In 1999, Hood co-founded the ISB, a nonprofit research institute focused on systems biology. From 2000 to 2020, researchers affiliated with ISB published 1,733 WOS-indexed articles.[10]

---

[10] Based on a search limited to "scientific articles" with an ISB affiliation conducted on October 15, 2020 using the bibliometrix package in R.

These articles have been cited more than 180,000 times. The top journals in which the ISB research appears are *Proceedings of the National Academy of Sciences of the United States of America* (80 articles), *Molecular & Cellular Proteomics* (69), and *PLOS One* (62). The ISB's scientific focus reflects its concentration on systems biology. The top author-provided keywords for ISB articles are "proteomics" (61 articles), "mass spectrometry" (39), and "systems biology" (30). The top collaborating organizations, as indicated by author affiliations, include the University of Washington (846 articles), the University of Texas MD Anderson Cancer Center (218), the University of Zurich (199), the Fred Hutchinson Cancer Research Center (179), and the University Michigan (167).

Figure 4.3 shows the annual output of scientific publications by ISB authors from 2000 to 2020.[11] The scientists with the most ISB-affiliated publications are Hood and Ulrike Kusebauch.

Figure 4.3. Scientific Publications of the Institute for Systems Biology, 2000–2020

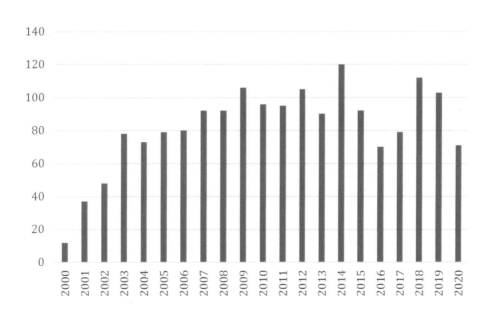

As described earlier, Hood's DNA sequencer was one of the key technologies that enabled the completion of the Human Genome Project in 2003. The International Human Genome Sequencing Consortium's initial sequencing and findings cites the DNA sequencer in the "Technology for Large-Scale Sequencing" section of the report (Lander et al., 2001). The publication has been cited more than 14,000 times in WOS-indexed articles and more than 25,000 times according to Google Scholar.

---

[11] Publications are reported as of October 15, 2020; therefore, the 2020 data may be incomplete.

Technological Impact

As of January 1, 2020, Hood has been named as an inventor on 72 patents. These patents have been cited nearly 1,400 times by subsequent patents by 309 entities, including individuals, corporations, and institutions, and their divisions, subsidiaries, and affiliates. Table 4.2 summarizes Hood's technological impact using the primary measures of technological output used throughout this report.

Table 4.2. Summary of Hood's Technological Impact

Total patents	72
Forward citations	1,377
Citing organizations	309

Although Hood's most widely recognized and frequently cited patents were granted earlier in his career, 2005 and 2006 were Hood's most-active years in terms of patent output.[12] Hood is a co-inventor on dozens of patents with several notable scientists, including Edward Jung, Clarence Tegreene, Lowell Wood, Jr., Muriel Ishikawa, Robert Langer (another Lemelson-MIT Prize winner), and Victoria Wood.

Hood's patent portfolio spans a wider variety of disciplines than many of his peers, reflecting the interdisciplinary nature of his research. Figure 4.4 provides the distribution of Hood's patent portfolio by WIPO technology sector. About half of Hood's patents are classified under chemistry, 38 percent are classified under instruments, and 9 percent are classified under electrical engineering.

---

[12] Based on priority year (the year in which the first application in a patent family is filed).

Figure 4.4. Distribution of Hood's Patents by Technology Sector

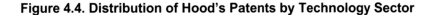

Figure 4.4. Distribution of Hood's Patents by Technology Sector

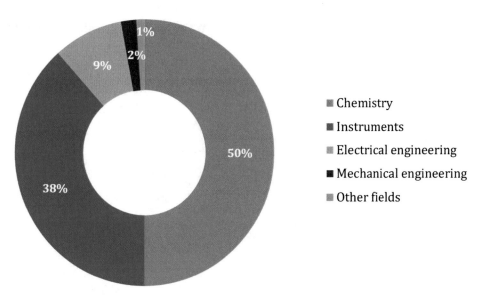

SOURCE: RAND analysis of patent data provided by IPVision, Inc.
NOTE: WIPO technology sector determined by IPC-WIPO sector crosswalk (Schmoch, 2008)

As described in Chapter 2, a patent's forward citations refer to instances when the patent is cited by subsequent patenting. Because patent applicants are required to cite, as prior art, all patents that were instrumental to the invention in question, forward citations thus represent a measure of technological impact. The technological impact of Hood's inventions is reflected in the forward citations of future patent applications to his patent portfolio. The organizations with the most forward citations to Hood's inventions are primarily biotechnology firms, but also include the United States and academic institutions, such as Cornell University and the University of California, Santa Barbara. The major biotechnology firms with the most forward citations to Hood's inventions are Pacific Biosciences of California, Life Technologies (a merger of Invitrogen Corporation and Applied Biosystems Inc., later acquired by Thermo Fisher Scientific), Affymetrix (acquired by Thermo Fisher Scientific), Illumina, Goldman Sachs,[13] and Applied Biosystems (which merged with Invitrogen Corporation and later was acquired by Thermo Fisher Scientific). The separate appendix provides a summary of the top citing organizations to Hood's patent portfolio.

Table 4.3 provides examples of new inventions among the top 25 citing organizations (i.e., entities with the highest number of forward citations) to Hood's patent portfolio that would, in

---

[13] Goldman Sachs Bank USA is the current assignee for 50 patents with forward citations to Hood. It likely took ownership of these patents when it financed the acquisition of Gen-Probe Incorporated by Hologic, Inc. in 2012.

turn, lead to new discoveries and patents, each having 100 or more forward citations themselves.[14]

Table 4.3. Notable Inventions with Forward Citations from Hood's Patent Portfolio

Citing Organization	Total Patents with Forward Citations	Examples of Notable Inventions	Year
Pacific Biosciences of California	70	Fluorescent nucleotide analogs and uses thereof	2008
Life Technologies	66	Molecular motors	2001
		Methods and products for analyzing polymers	2002
Affymetrix	60	Very large-scale immobilized polymer synthesis using mechanically directed flow paths	1995
		Combinatorial strategies for polymer synthesis	1997
Illumina	53	Detection of nucleic acid amplification reactions using bead arrays	2002
		Method of making and decoding of array sensors with microspheres	2006
Goldman Sachs Bank USA	50	Detection of nucleic acid sequences by invader-directed cleavage	1999
		Invasive cleavage of nucleic acids	1999
Applied Biosystems	49	Coupled amplification and ligation method	1999
Cornell University	43	Detection of nucleic acid sequence differences using coupled ligase detection and polymerase chain reactions	2000
		Detection of nucleic acid sequence differences using the ligase detection reaction with addressable arrays	2003
Theranos	26	Systems and methods for multi-analysis	2013
University of California, Santa Barbara	16	Nucleic acid detection methods using universal priming	2004
United States of America	15	High-speed parallel molecular nucleic acid sequencing	2006
		Method for detecting the presence of a single target nucleic acid in a sample	2011
Agilent Technologies	13	Methods and compositions for determining the sequence of nucleic acid molecules	2001
		Light source power modulation for use with chemical and biochemical analysis	2002

SOURCE: RAND analysis of data provided by IPVision, Inc.

---

[14] The inclusion of Theranos (a health technology company that went defunct after its technology was proven to be based on fraudulent claims) in Table 4.3 highlights one of the limitations of looking at forward citations in isolation. We do not know if the Theranos patents that cited Hood's patents were part of Theranos' falsified scientific evidence.

Economic and Social Impacts

Hood laid the technological foundation for genomics and proteomics through the invention of cutting-edge scientific instruments incorporating automation with integrated laser and computer technology. These instruments allowed rapid automated protein and DNA sequencing and led to significant advancements in mapping and sequencing human genes and discovery of new proteins leading to improved medical treatments. Hood was affiliated with multiple universities and organizations, and he co-founded more than a dozen companies throughout his career. Hood's later pioneering research in systems biology and systems medicine would lead to advances in entirely new fields, and those contributions are still being realized today.

California Institute of Technology

From 1970 to 1992, Hood was a professor of biology at Caltech. During this time, Hood and his colleagues pioneered the first sequencing machines. One of the Caltech researchers' significant contributions was the development of fluorescent dyes to replace radioactive labels, increasing the accuracy and safety of DNA sequencing and helping to automate procedures. Hood's patents on the automated DNA sequencer and other instruments were owned by Caltech and exclusively licensed to Applied Biosystems for commercialization. The *New York Times* estimated that Caltech earned tens of millions of dollars from the DNA sequencer and related patents, although the attribution of credit became the subject of controversy because not all of the researchers in Hood's lab were named as inventors on the patents (Grosselin and Jacobs, 2000; Pollack, 2016).

As evidenced by the nature of his inventions, Hood was a leading proponent of cross-disciplinary research. In 1989, Hood stepped down as chairman of the Division of Biology to become the director of the Center for Molecular Biotechnology at Caltech, a new science and technology center funded by the National Science Foundation. Around this time, Hood had more than 100 researchers working in his lab, much larger than similar groups at Caltech. The center funded research and development for new and improved instruments, including for protein sequencing and large-scale sequencing of genomic DNA. The center also funded education and outreach programs to provide research experience and training for high school, undergraduate, graduate, and postdoctoral students (National Science Foundation, undated). These included programs to encourage increased participation of underrepresented minority groups in scientific research.

The Human Genome Project

The DNA sequencer was one of the technologies that enabled the Human Genome Project, the largest international collaboration ever undertaken in the field of biology. This project required billions of dollars in government funding, and the research capabilities of thousands of scientists (Timmerman, 2016). Hood's lab at Caltech, with funding from the U.S. Department of

Energy and the National Institutes of Health, became one of dozens of research centers in the United States, Europe, and Asia that were sequencing the human genome. The project began in 1990 and was completed in 2003. The International Human Genome Sequencing Consortium (2004) published a scientific description of the estimated 20,000 to 25,000 human genes to make them accessible to researchers for further biological study.

The Battelle Memorial Institute (2013) estimated that the Human Genome Project had contributed more than $1 trillion in direct and indirect economic activity, including supporting further scientific research after its completion. The U.S. government invested $3.8 billion (approximately $6.1 billion in 2019 dollars) in the project through its completion and subsequently funded $8.5 billion (approximately $10.2 billion in 2019 dollars) in related research and funding support. This suggests that every $1 in initial federal funding for the project contributed an additional $178 in economic output in the United States (an additional $65 after accounting for all federal funding for human genome related research).[15] Other researchers noted that the Battelle estimate failed to account for the impact of private investment and technological innovation, which also contributed to the subsequent growth in practical applications of genomics (Wadman, 2013). Economists generally believe that the Battelle estimate is an imperfect measure of the impacts of health research, as it does not include improvements in human health caused by better health outcomes and the development of new drugs and diagnostics (Hall, Mairesse, and Mohnen, 2009). The $1 trillion estimate may be misleading because, although it represents the level of economic activity engaged in mapping and sequencing the human genome and its applications, it does not directly measure outcomes and it implies that funding for scientific research; jobs in science, technology, and engineering; and researcher capabilities would not have otherwise been productively employed in other areas of research. This does not diminish the many scientific and medical accomplishments the project did achieve, which include the mapping of the number, location, size, and sequence of human genes; the production of specific gene probes to detect carriers of genetic diseases; the discovery of new proteins leading to many improved medical treatments; and the comparison of different genomes to provide insights into human origins and ancestry.

Applied Biosystems

While at Caltech, Hood tried to sell his instruments to more than a dozen companies before deciding to start his own company with the help of venture capitalist William Bowes, a member of the board of directors of Cetus Corporation from 1972 to 1978. Applied Biosystems, Incorporated, initially Genetic Systems Company (GeneCo), was founded in 1981 in Foster City, California, to commercialize the technology developed by Hood and his colleagues. Its first commercial instrument, the Model 470A Protein Sequencer, was released in late 1982. The

---

[15] These estimates were developed using IMPLAN's economic analysis tool for U.S. economic impacts; the tool uses an input-output framework (Battelle Memorial Institute, 2013).

company's first-year sales were approximately $400,000, and it had about 40 employees. In 1983, Applied Biosystems released its second commercial instrument, the Model 380A DNA Synthesizer, and was listed on the NASDAQ exchange with revenues of $5.9 million (Applied Biosystems, undated).

In 1986, Applied Biosystems introduced the first commercial DNA sequencer. The promise of the new technology would motivate the U.S. government and scientists around the world to initiate research on the Human Genome Project. A few firms introduced competing sequencing instruments, but none replicated Caltech's fluorescent dye technology. The technology improved, and by 1999, a fully automated DNA sequencer could sequence up to 150 million base pairs per year—significantly advancing the complete mapping and sequencing of the human genome.

By 2007, Applied Biosystems had approximately 1,000 employees in eight countries and $132 million in annual revenues, with more than 25 different automated instruments and hundreds of consumables, chemicals, and components (BioSpectrum, 2007). Table 4.4 describes some of the early instruments developed by Applied Biosystems and their impacts.

Table 4.4. Applied Biosystems' Early Products and Impacts, 1982–1986

Year	Product	Purpose	Impacts
1982	Model 470A Protein Sequencer	Instrument used to identify the amino acid sequence within a purified protein	Contributed to more than 2,000 scientific publications, including sequencing the Jack Bean Urease, luminescent protein aequorin from a jellyfish, skeletal muscle sarcoplasmic reticulum from a rabbit, and TGF-β in humans
1983	Model 380A DNA Synthesizer	Instrument used to automate the phosphonamidite chemistry method for DNA synthesis	• Automated production of high-quality oligonucleotides, which contributed to the invention of the PCR and other methods for genetic disease detection • Reduced time and cost in the scientific discovery process through automation
1984	Model 430A Peptide Synthesizer	Protein synthesizer capable of building long peptides using amino acids	Allowed researchers to produce large volumes of proteins, facilitating large-scale experimentation at a reduced cost
1986	Model 370A DNA Sequencing System	Instrument used to automate the Sanger DNA sequencing procedure; first commercial DNA sequencer	• Contributed to development of expressed sequence tags, a revolutionary new method for gene discovery and sequencing • One of the key technologies that enabled the completion of the Human Genome Project
	Model 340A Nucleic Acid Extractor	Instrument used to automate isolation of DNA or ribonucleic acid (RNA) from tissue or cells in culture	• Produced high yields of high-quality genomic DNA and RNA from a variety of environmental samples • Allowed medical labs to isolate DNA from bacteria, blood, and tissue

In 1993, Applied Biosystems was acquired by the Perkin-Elmer Corporation and became the Applied Biosystems Division (later PE Applied Biosystems and PE Biosystems) in a deal valued at approximately $330 million. At the time, Perkin-Elmer was the world's largest manufacturer of instruments and reagents for PCR, for which it marketed reagents kits with Hoffman-La Roche. The Applied Biosystems Division sold automated sequencers and thermal cyclers that bolstered the nascent genomics industry in developing new pharmaceuticals based on the work of the Human Genome Project (Perkin-Elmer, 1994).

In 1998, PE Biosystems' president, Michael Hunkapiller (of Caltech), made an ambitious bet that private industry could decode the human genome faster than the academic consortium. Because of Applied Biosystems' close ties to the Human Genome Project, PE Corporation created Celera Corporation as a separate business unit led by Craig Venter, the former president of the nonprofit Institute for Genomic Research, as one of several independent competitors to the International Human Genome Sequencing Consortium (Shreeve, 2007). Celera was formed to commercialize genomic information; it generated controversy by seeking to sequence sections of the human genome for commercial gain and rejecting the open-access policy of the Human Genome project, a decision it would later rescind. Celera sequenced the human genome at a fraction of the cost of that of the Human Genome project; however, a significant portion of the human genome had already been sequenced, and Celera benefited from using the consortium's existing data from the freely available, open-access GenBank database developed and maintained by the National Institutes of Health. Nonetheless, the introduction of privately funded competitors who benefited from the initial work of the consortium may have accelerated the publicly funded effort to sequence the entire human genome, which was finished two years ahead of schedule.

In 2000, PE Corporation changed its name to Applera, an amalgamation of Applied and Celera Corporation, and once again began using the Applied Biosystems Group name. In 2008, Applied Biosystems merged with Invitrogen in a deal valued at approximately $6.7 billion, forming Life Technologies with combined sales of about $3.5 billion, 9,500 employees, and more than 3,600 patents and exclusive licenses ("Applied Biosystems, Invitrogen Complete $6.7 Billion Merger," 2008). Life Technologies was acquired by Thermo Fisher Scientific in 2014 for approximately $13.6 billion (Berkrot and Kelly, 2013).

Amgen

In 1980, William Bowes recruited George Rathmann from Abbott Laboratories and Winston Salser from the University of California, Los Angeles, to found Amgen (originally Applied Molecular Genetics Inc.) in Thousand Oaks, California. Amgen had a scientific advisory board consisting of Hood, Marvin Caruthers, and other prominent scientists. Rathmann, the founding chief executive officer (CEO), who became interested in rDNA technology at Abbott Laboratories, said that Hood "accelerated the whole field of biotechnology by a number of years" (Timmerman, 2016). Amgen would focus on molecular biology and biochemistry to develop health care treatments using rDNA technology. On June 17, 1983, Amgen's IPO raised nearly $40 million (Amgen, undated).

In the early 1980s, Fu-Kuen Lin, an Amgen research scientist from Taiwan, and his team sought to find and clone the erythropoietin gene from a hormone that had been isolated by Eugene Goldwasser at the University of Chicago. Hood's protein sequencer was used to sequence erythropoietin. The two-year Amgen project led to the development of Epogen, a treatment for anemia in chemotherapy patients. Epogen is one of the most successful drugs in the biotechnology industry, and has been prescribed to more than 1.5 million Medicare patients since obtaining FDA approval in 1989 (Epogen, undated). In 1984, Amgen formed a joint venture with Kyowa Hakko Kirin, a Japanese pharmaceutical and biotechnology company, for the global development of Epogen. Amgen received the first patent for recombinant human erythropoietin in 1988, and Epogen was approved by the U.S. FDA in June 1989. Epogen generated approximately $150 million in sales in the ten months after its launch. From 2003 to 2010, Epogen sales peaked at around $2.5 billion per year; in 2019, sales were slightly below $900 million (Amgen, 1995–2019).[16]

---

[16] These data are derived from Amgen, 1993–2019.

In 1985, Larry Souza and his team cloned granulocyte colony-stimulating factor (G-CSF), leading to the development of Amgen's second major drug, Neupogen, for the prevention of infections in immunocompromised patients. Neupogen was approved by the FDA in February 1991. Epogen and Neupogen were both named "Product of the Year" by *Fortune* magazine, for 1989 and 1991, respectively (Amgen, undated). In 1992, Amgen reached $1 billion in sales for the two drugs and was added to the Standard & Poor's 500 Index. Over the next few decades, Amgen developed more than a dozen new drugs and acquired several biotechnology and pharmaceutical companies and products. Figure 4.5 summarizes Amgen's product sales by year.

Figure 4.5. Amgen Product Sales by Year

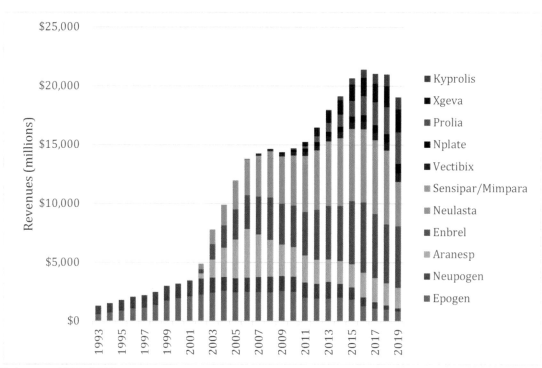

SOURCE: Amgen, 1995–2019.

In 2019, Amgen had about 22,000 employees and revenues of $23.4 billion, and invested more than $4 billion in new research and development (Amgen, 2020). Amgen's market capitalization was $142.6 billion as of December 31, 2019.[17] Amgen's top-selling drugs in 2019 were Enbrel, Neulasta, Prolia, Xgeva, Aranesp, Kyprolis, and Epogen. Amgen also reported having more than 30,000 patients enrolled in some 150 clinical trials for potential new medicines (Amgen, 2020). Table 4.5 describes major products developed by Amgen and their impacts.

---

[17] Search of Yahoo! Finance, Valuation Measures, for Amgen Inc. conducted on December 18, 2020.

Table 4.5. Amgen's Major Products and Impacts

Drug or Product	Year Approved	Purpose	Impacts
Epogen	1989	Treatment of anemia associated with chronic renal failure, including patients on dialysis and patients not on dialysis; later approved to treat anemia due to chemotherapy and treatment with certain HIV drugs	• Decreased the need for blood transfusions due to increased hemoglobin levels • Patient-reported improvement in overall physical function
Neupogen	1991	Prevention of infections in immunocompromised patients undergoing chemotherapy	Reduced risk of infection and infection-related mortality in chemotherapy patients
Enbrel	1998	Treatment of inflammatory conditions, including moderate to severe rheumatoid arthritis and moderate to severe plaque psoriasis	• Reduced levels of pain and fatigue • Delayed progression of joint damage
Aranesp	2001	Treatment of anemia associated with chronic renal failure, including patients on dialysis and patients not on dialysis	Allowed erythropoietin (see Epogen) to remain in the body longer
Neulasta	2002	Treatment of neutropenia, a lack of certain white blood cells in patients undergoing chemotherapy, that stimulates the growth of white blood cells	Reduced risk of infection and infection-related mortality in chemotherapy patients
Sensipar/Mimpara	2004	Treatment for increased level of parathyroid hormone in patients with long-term kidney disease who are on dialysis	Prevention of bone disease by decreasing the amount of parathyroid hormone, calcium, and phosphorus in the body
Vectibix	2006	Treatment for colorectal cancers that have spread or worsened during or following chemotherapy	Delayed progression of metastatic colorectal cancers
Prolia	2010	Treatment of post-menopausal osteoporosis	Reduced risk of vertebral fractures and hip fractures
Xgeva	2010	Prevention of skeletal-related events in advanced cancer patients who have bone damage	Reduced pain levels and reduced risk of bone fractures
Kyprolis	2012	Treatment of patients with relapsed or refractory multiple myeloma	• Delayed progression of multiple myeloma • Reduced treatment days due to improved efficacy

NOTE: All drug treatments pose the risk of serious side effects that must be weighed against the potential benefits of treatment.

In addition to becoming one of the most successful biotechnology companies, Amgen endeavored to make a positive social impact. In 1994, Amgen became the first biotech company to receive a National Medal of Technology from the U.S. Department of Commerce for "its leadership in developing innovative and important cost-effective therapeutics based on advances in the cellular and molecular biology for delivery to critically ill patients throughout the world" (Amgen, undated). In 2006, Amgen collaborated with the Brigham and Women's Hospital and the National Institutes of Health's National Heart, Lung, and Blood Institute on the Women's

Genome Health Study to identify genetic variations that may contribute to a variety of illnesses, including breast cancer, diabetes, heart disease, and osteoporosis. In 2007, the Amgen Foundation, in collaboration with MIT, launched the Amgen Scholars program to provide undergraduates with access to research experiences and exposure to biotechnology and drug discovery at top institutions globally (Amgen, undated).

Institute for Systems Biology

In December 1999, Hood left the University of Washington to co-found ISB, a nonprofit collaborative and cross-disciplinary biomedical research organization, with Alan Aderem, an American biologist specializing in immunology and cell biology, and Reudi Aebersold, a pioneer in the fields of proteomics. Based in Seattle, Washington, ISB has funded groundbreaking cross-disciplinary research in human health issues, including aging, brain health, cancer, and chronic and infectious disease. ISB has fostered scientific collaboration between academia and industry, producing more than a thousand scientific publications (ISB, undated-b) and leading major clinical studies (ISB, undated-c).[18] ISB is an affiliate of Providence St. Joseph Health, one of the nation's largest nonprofit health care systems. ISB is also involved in many philanthropic efforts. In 2018, with more than 200 employees, ISB received nearly $22 million in grants and contract revenue and $3.4 million in contributions (ISB, undated-a). While ISB has already made significant scientific contributions, many of its economic and social impacts are yet to be seen because the institution is conducting early-stage research in several emerging fields, including systems genetics, scientific wellness, and sustainable agriculture. Table 4.6 describes some of ISB's research contributions and early and potential longer-term impacts.

---

[18] See additional discussion of ISB's scientific publications and collaborations the Scientific Impact section earlier in this chapter.

Table 4.6. Institute for Systems Biology Economic and Social Impacts

Topic	Research/Project	Purpose	Impacts
Cancer	Regulatory network for glioblastoma using systems genetics network analysis (published in *Cell Systems*)	Developed a platform for integrating somatic mutations and gene expression from patient data to construct the most comprehensive regulatory network for the deadly brain cancer glioblastoma multiforme	Demonstrated that clinical and genomic data could be used to develop a regulatory network to gain meaningful clinical and biological insights
Genetics	High-resolution mapping of Huntington's disease in mice (published in *Human Molecular Genetics*)	A multi-institute collaboration to map in high resolution the earliest effects of the Huntington's disease mutation in mice	Mapping early pathogenesis expanded basic understanding of early disease transitions and will inform other studies on proximal mechanisms and how to slow the disease process
Proteomics	The Human SRMAtlas	To provide mass spectrometry assays based on selected reaction monitoring (SRM) for essentially every human protein	Protein assays can be used in biological studies to identify and quantify any human protein and navigate complete proteome maps to understand their biological function
Sustainable Agriculture	Project Feed 1010	To standardize, automate, and democratize aquaponics systems, which can recycle water and nutrients to support year-round crop production in different environments	STEM curriculum devoted to climate change, food security, and environmental sustainability
Vaccine Research	Proteogenomic analysis of the malaria parasite *Plasmodium vivax* (published in *PLOS Neglected Tropical Diseases*)	An international collaboration to identify proteins in the malaria parasite using mass spectrometry	Research aims to provide new targets for a malaria vaccine
Wellness	Pioneer 100 Wellness Project	Tracked more than 100 study participants using whole genome sequencing, blood and urine analysis, and activity tracking to create personal, dense, dynamic data clouds to optimize individual wellness	• Study established thousands of statistical correlations in data and identified novel insights into human biology • Personal data clouds can identify early transitions into disease states and facilitate return to wellness with behavioral coaching

Involvement in Other Organizations

Hood co-founded more than a dozen companies and has served on more than three dozen corporate advisory boards. Hood is also one of only about a dozen members elected to all three national academies: the National Academies of Sciences, Medicine, and Engineering.

After co-founding Applied Biosystems and Amgen, in 1987, Hood helped launch SyStemix, formed by Irv Weissman of Stanford University, to develop blood-marrow cell and gene therapies. In 1991, Weissman sold a majority share to Sandoz Pharma, a Swiss drug and chemical company, for $392 million. Sandoz later merged with Ciba-Geigy to become Novartis, which bought the remainder of SyStemix (and its patents) for $76 million in 1997. SyStemix's pioneering work in stem cell research was shut down by Novartis in 2000. Weismann was able to license key technologies from Novartis, and more than a decade later, resumed research to grow and deliver blood stem cells to cancer patients, boosting survival rates (Krieger, 2016).

In the early 1990s, Hood also helped launch companies attempting to commercialize the benefits of increasing computing power to use genetic data and text analysis systems and to search for compounds with pharmaceutical promise. In 1992, Darwin Molecular was founded as a genomics company with other notable investors, including Microsoft founders Bill Gates and Paul Allen. The company was acquired by Chiroscience Group in 1996 for $120 million and later by Celltech Group in 1999. It ceased operations in 2003. Hood also helped found Paracel in 1992; it developed supercomputers to conduct text searches and compare letters in gene sequences. Paracel was acquired by Celera Genomics Group in 2000 for $245 million (Bell, 2000).

In the mid-to-late 1990s, Hood also co-founded Prolinx (a biotechnology company that developed tools, systems, and applications to advance developments in proteomics and genomics by facilitating the manipulation of macromolecules), Rosetta Inpharmatics (a manufacturer of DNA microarray gene expression systems to identify gene functions and drug targets to advance drug discovery), and T Cell Sciences (a manufacturer of pharmaceutical products using proprietary T-cell receptor and soluble receptor technology to monitor immune system disorders). Prolinx ceased operations in 2003. Rosetta was acquired by Merck in 2001 for $620 million and ceased operations in 2008. T Cell Sciences acquired the Virus Research Institute in 1998 for $70 million; the combined company changed its name to Avant Immunotherapeutics, which merged with Celldex Therapeutics in 2008 (Philippidis, 2014).

At the University of Washington, Hood invented the ink-jet DNA synthesizer, which could create DNA arrays with tens of thousands of gene fragments. Agilent Technologies, which was spun out from Hewlett-Package in 1999, commercialized the ink jet DNA synthesizer. Agilent's 1999 initial public stock offering raised $2.1 billion and set a record at the time as the largest IPO in Silicon Valley. Agilent became a leading manufacturer of analytical and scientific instruments. Agilent employed approximately 48,000 people in 2001, before slowing sales and economic uncertainty forced the company to downsize and sell off certain divisions, including its health care and medical products division and its semiconductor business. In 2019, Agilent had 16,048 employees and revenues of $5.16 billion (Agilent Technologies, 2019).

In 2000, Hood co-founded Macrogenics (a biotechnology company that developed immune-therapeutics for treating cancer and inflammatory disorders) and Phenogenomics Corporation (a biotechnology company seeking to develop rapid methods for detecting proteomic interactions designed to discover new antibody-based drugs for cancer treatment) (Agilent Technologies, 2019).

In 2003, Hood co-founded and served on the board of directors of Accelerator Corporation (later Accelerator Life Science Partners), an investment company that provides venture capital funding and management for emerging biotechnology companies. Accelerator has funded about 20 companies with more than $165 million in equity financing (Accelerator Life Science Partners, undated). In 2009, Hood also helped launch and served on the board of directors of Integrated Diagnostics, a manufacturer of diagnostics and measurement technologies capable of monitoring hundreds of biomarkers simultaneously for earlier detection and more-accurate management of complex diseases, including lung cancer and Alzheimer's disease. Integrated Diagnostics developed the XL2 test, a noninvasive blood test that measures proteins to identify lung nodules that have a high probability of being benign. Every year, health care providers in the United States identify more than 1.5 million lung nodules in patients (Gould et al., 2015). Integrated Diagnostics was acquired by Biodesix, a molecular diagnostics company, in 2018.

Hood was a co-founder, investor, or adviser in more than a dozen companies (the largest being Amgen, Agilent Technologies, and Applied Biosystems) that collectively created more than 40,000 jobs, generated more than $30 billion in annual revenues, and/or were involved in mergers or acquisitions with other organizations in market transactions valued at approximately $2.5 billion in 2019 dollars.[19]

Conclusion

Hood's scientific, technological, economic, and social impacts owe largely to his contributions in the advancement of scientific instrumentation. Hood has authored hundreds of scientific publications that have been cited thousands of times and been granted dozens of patents, but thousands of scientists in a diverse array of fields, including genomics, proteomics, molecular biology, oncology, and neuroscience, have used his instruments. Hood has continued to innovate, pioneering research in entirely new fields, such as systems biology and systems medicine.

Hood's instruments have been used, among other things, to sequence the prion protein (leading to a better understanding of fatal neurodegenerative diseases), synthesize oligonucleotides (enabling the invention of the polymerase chain reaction), synthesize the 99 residue HIV-1 protease (leading to the development of the protease inhibitor to treat HIV/AIDS), and enable the mapping and sequencing of the entire human genome. The Human Genome Project, for which Hood was an early advocate and key contributor, was the largest international

---

[19] Based on analysis of Yahoo! Finance, Dun & Bradstreet, and other financial news sources. Note that in Chapter 2 where we reported the collective impacts of all of the Lemelson-MIT Prize winners, we only included economic and financial data for companies founded by the inventors, but not other companies where they served as advisers, investors, or in other capacities.

collaboration ever undertaken in the field of biology, involving thousands of scientists and billions of dollars of investment by the U.S. government and others.

Hood was a co-founder, investor, or adviser in more than a dozen companies that collectively provide more than 40,000 jobs and generate more than $30 billion in annual revenues. Several of these companies combined, through mergers or acquisitions, with other organizations to achieve significant advances in genetic research, pharmaceutical manufacturing, and medicine. The impact and contributions of Hood's later founding of ISB and his pioneering research in entirely new fields of study are still being realized today.

5. Bertozzi: Glycoscience and Bioorthogonal Chemistry

Recent advancements in the understanding of chemistry, biology, and the intersections of the two fields are improving humanity's ability to discover and create new and more effective treatments and diagnostics for a variety of diseases. It is now understood that the surface of human cells is not a simple smooth barrier—cell surfaces are coated with molecules of a wide variety of shapes and sizes. These coatings, which are made of *glycans*—sugar molecules[1]—affect how the cell interacts with its surrounding environment and may even reflect the status of the cell itself (Arizona State University, 2017). *Glycoscience* involves the study of these molecules, such as from a chemical or biological perspective. Glycoscience research is helping develop new medicines that directly interact with these coatings in a variety of ways to identify, treat, or even potentially cure diseases.

Carolyn Bertozzi, the Anne T. and Robert M. Bass Professor of Chemistry and Professor of Chemical and Systems Biology and Radiology (by courtesy) at Stanford University and an Investigator of the Howard Hughes Medical Institute, has been a vanguard of this research and development. Her work studies the chemistry of cell surfaces, with a focus on how these surfaces vary because of such diseases as cancer. Bertozzi has sought not only to improve humanity's understanding of the reactions involving of glycans on both healthy and diseased cells but also to develop new methods and technologies for interacting with and exploiting these chemical reactions for the purpose of developing new medical diagnostics and treatments.

These efforts have seen Bertozzi credited with helping found an entirely new field: *bioorthogonal chemistry*. As the most recent Lemelson-MIT Prize winner of the three case studies examined in this report, her inventions are still in their infancy, and the full scale of her work's impact is yet to be realized. Nevertheless, Bertozzi has already published hundreds of scientific articles, developed more than 60 patents, and co-founded seven companies.

Overview of Inventions

Bioorthogonal chemistry, the field Bertozzi is credited with helping found, is only a handful of decades old. "Bioorthogonal reactions are chemical reactions that neither interact with nor interfere with a biological system" (Sletten and Bertozzi, 2011). Traditionally, chemists have developed reactions that occur in controlled environments. In these controlled environments, chemists can precisely manage the temperature, pressure, and presence or absence of other molecules to ensure an introduced reagent interacts with another chemical substance—and only

---

[1] *Sugar* as a chemical term refers to all carbohydrates of the general formula $C_n(H_2O)_n$, not just those which are used in food. Sugar molecules and combinations of sugar molecules come in a wide variety of forms and are a basic building block for many biological processes.

that chemical substance—in the desired way. Hundreds of years of this type of chemistry have vastly improved humanity's understanding of atomic and molecular structures, the nature of chemical reactions, and how to design and control the desired synthetic reaction. Bertozzi explains that, as a result of these years of research,

> [t]he target-driven synthetic chemist now enjoys an impressive reaction toolkit. But what if the challenge were inverted, wherein the target structure was relatively simple but the environment in which the necessary reactions must proceed was so chemically complex and uncontrollable that no two functional groups could combine reliably and selectively under such conditions? (Bertozzi, 2011)

The challenge faced by bioorthogonal chemistry is to identify chemical reactions that can reliably take place in more-complex environments, with the ultimate goal of identifying reactions that can reliably and safely occur within live animals—including, ultimately, humans. Figure 5.1 illustrates this complexity and how it relates to Bertozzi's field of glycoscience. Figure 5.1 illustrates the challenge of identifying a molecule, Y, that reliably and safely reacts with X to form X-Y—and does so without reacting to any of the other molecules that it may encounter on its way to X.

Figure 5.1. An Illustration of a Generic Bioorthogonal Reaction, X+Y → X-Y

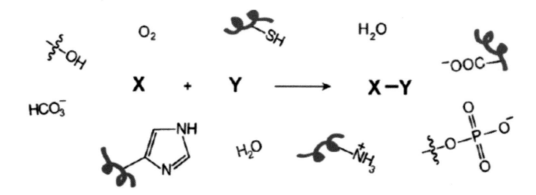

SOURCE: Sletten and Bertozzi, 2011, p. 667. Used with permission. Permission requests related to this figure should be directed to the American Chemical Society.
NOTE: This figure shows a molecule, Y, that interacts with another molecule, X, without interacting with the many other molecules in its environment.

Until recently, there were no known chemical reactions capable of meeting these criteria. "A few isolated reports from the 1990s suggested that such chemical reactions might exist or [at] least could be invented with some clever mechanistic thinking." (Sletten and Bertozzi, 2011). In 2003, Bertozzi's team published the first use of the term bioorthogonal (Hang et al., 2003). Since that time, interest in bioorthogonal chemistry has been exploding. As Devaraj (2018) notes, "The concept is elegant. Can we design reactions that are so selective they can be performed reliably even in a complex biological environment?" Several additional bioorthogonal reactions have been introduced since that time, by both Bertozzi and others.

Identifying a bioorthogonal reaction is not simple. Sletten and Bertozzi (2011) describe the process: Once a chemical reaction with properties that suggest it may have potential to operate in a biological environment—a challenge in itself—has been identified, a researcher must test "whether the reaction proceeds reliably in aqueous media alongside biological metabolites such as amino acids and sugars," then test in increasingly complex biological environments, such as "live cells, and ultimately, in model organisms such as zebrafish or mice." This iterative process often requires creative modification to the chemical structures involved in the reaction. Further, it is not sufficient for the reaction to simply take place in these settings without disruption, as the goal is to eventually enact some change in the biological system. Therefore, at least one of the reaction's participating functional groups can react with the biomolecules of living systems.

Devaraj (2018) describes a wide variety of valuable medical applications that could eventually be enabled by bioorthogonal chemistry. One application, following Bertozzi's foundational Hang et al. (2003) publication and interest in cell surface glycans, is to use bioorthogonal reactions for the formation of *bioorthogonal handles*. Figure 5.2 illustrates the potential medical value of such a reaction. If X is one of the many glycans that coat the outside of a cell or reside within a cell, a bioorthogonal reaction with Y could be used to ensure Y—and anything attached to Y—is brought directly to cells that have the glycan X. If X is a glycan that is specific to cells with a certain disease, such as cancer, a bioorthogonal reaction could be used to flag diseased cells for destruction by medicines or even by a body's natural defenses.

Figure 5.2. An Illustration of a Generic Bioorthogonal Reaction, X+Y → X–Y, as a Bioorthogonal Handle

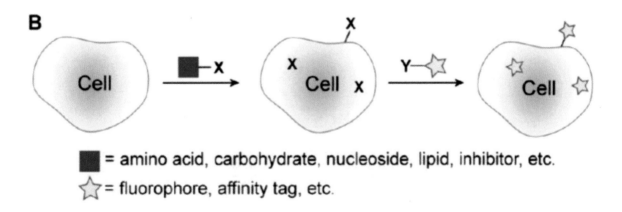

SOURCE: Sletten and Bertozzi, 2011, p. 667. Used with permission. Permission requests related to this figure should be directed to the American Chemical Society.

Another potential application of bioorthogonal chemistry would be to enable "applications that involve the assembly of bioactive small molecules in situ," potentially even "in situ assembly of pharmaceuticals at the site of their action" (Devaraj, 2018). In other words, a bioorthogonal reaction could be used to assemble a medically effective compound at the site of disease, rather than trying to direct a prefabricated medical compound to the right location. This concept of on-site assembly is particularly appealing in cases in which the full compound should be delivered inside a cell or other area but is too large to pass through associated membranes or other barriers; it would also be useful in cases where precise targeting of the compound is very important. The assemblage of on-site compounds need not be limited to synthetic pharmaceuticals. Bioorthogonal reactions could, in theory, also be used to create natural compounds at specific locations that would then ideally proceed with their intended biological reactions—a novel approach for addressing diseases caused by the lack of normally present compounds.

Yet another potential application is "to uncage substrates, a strategy that has been termed 'click to release'" (Devaraj, 2018). This approach uses a bioorthogonal reaction not to bring together pieces of a new compound but to release a drug, imaging agent, or protein that is already present in a large molecule by reacting with the larger molecule in a way that causes it to release the desired drug, imaging agent, or protein. An illustration of this process is shown in Figure 5.3.

Figure 5.3. General Example of Bioorthogonal "Click to Release" Reaction

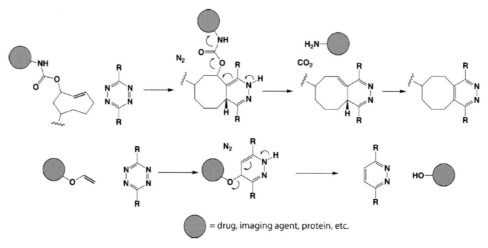

Figure 3. General examples of bioorthogonal uncaging reactions. Dienophiles act to mask functional groups such as amines and alcohols. Upon bioorthogonal reaction with tetrazine, the functional group is released. This enables the uncaging of drugs, imaging agents, and even enzymes.

SOURCE: Devaraj, 2018, p. 955. Used with permission. Permission requests related to this figure should be directed to the American Chemical Society.

Challenges remain in converting these possibilities into reality, particularly with respect to effectively and safely performing bioorthogonal reactions in live, multicellular organisms—as Devaraj (2018) says, "[although] performing bioorthogonal reactions on and within living cells has become almost a routine phenomenon, the ability to conduct highly efficient reactions inside multicellular organisms remains nontrivial." However, the challenges continue to be slowly and steadily overcome. Bertozzi and other researchers have successfully demonstrated applying bioorthogonal reactions in zebrafish embryos (Laughlin et al., 2008; Baskin et al., 2010; Jiang et al., 2011; Agarwal et al., 2015; Westerich et al., 2020), but tailoring bioorthogonal reactions to proceed efficiently in mammals remains a challenge. The reactions that worked well in zebrafish reacted less efficiently in mice (Sletten and Bertozzi, 2011). Even after these challenges are addressed, there will be further challenges in converting bioorthogonal reactions into medical products, such as optimizing "the dosing, pharmacokinetics, stability, and toxicity . . . for not one but two reagents." (Devaraj, 2018). However, Devaraj (2018) deems these issues as "likely surmountable."

For this reason, the ultimate impacts of Bertozzi's work remain to be seen. She has co-founded seven different companies based on different aspects of her research, but medical applications of her discoveries are still in trial phases, so the final outcome is uncertain. Nevertheless, the outlook is optimistic. "Ultimately, I imagine that bioorthogonal chemistry will go beyond research applications and preclinical testing and will be translated to applications in humans" (Devaraj, 2018). Although Bertozzi's discoveries have not yet been fully translated into human applications, Table 5.1 lists several applications that are currently under development at companies that Bertozzi co-founded.

Table 5.1. Ongoing Inventions—Applications of Bertozzi's Discoveries that Are Currently Under Development

Invention	Beneficiaries	Associated Organization
Site-specific SMARTag technology enables the generation of homogenous bioconjugates, engineered to improve performance and ease of manufacturing	Enables the identification of superior drugs from libraries of differentially designed conjugates, and the generation of homogenous bioconjugates engineered to enhance potency, safety and stability.	Redwood Bioscience Inc.
Antibody Detection by Agglutination PCR (ADAP), which couples specific DNA tags with antigens that are used to detect the presence of antibodies in your blood, saliva, or other bodily fluids	Patients benefit from advanced detection of disease; because ADAP is 1,000 to 10,000 times more sensitive than other tests, it can detect antibody activity for such diseases as HIV and Type I diabetes much sooner, allowing for treatment at earlier stages of the disease.	Enable Biosciences
Platforms that overcome the long-standing technical barriers that have delayed drug development in targeting the Siglec-Sialoglycan axis	Has the potential to activate both innate and adaptive anti-tumor immune responses and treat patients who do not respond to T cell checkpoint therapies.	Palleon Pharmaceuticals
A blood test that can detect ovarian cancer earlier	Individuals with ovarian cancer face significantly higher survival rates when the disease is detected earlier, as existing treatments are more effective in earlier stages of the disease.	InterVenn Biosciences
A dye, usable in a simple smear test, that fluoresces when incorporated in living mycobacteria, which cause tuberculosis	Allows tuberculosis, which ravages poor individuals in countries with minimal access to health care, to be detected rapidly, cheaply, and with high accuracy.	OliLux Biosciences
The first commercially available panel capable of assessing blood (serum or plasma) protein glycosylation in a site-specific manner	Enables research advances in such areas as oncology, autoimmunity, metabolism, and neurology.	InterVenn Biosciences
A drug discovery platform based on lysosomal biology; "In a 2019 publication, Bertozzi's team at Stanford demonstrated that a cation-independent receptor called CI-M6PR could be exploited to capture and drag extracellular proteins into cells, trafficking them to the lysosome for destruction" (Edelson, 2020).	The goal is to discover and develop "first-in-class therapeutics that degrade extracellular and membrane-bound proteins that drive a range of difficult-to-treat diseases, including cancers and autoimmune conditions." (Versant Ventures, undated)	Lycia Therapeutics

Despite the remaining uncertainty, the impact of bioorthogonal chemistry has already been hailed as "monumental. Suddenly, reactions that previous generations performed in refluxing toluene, were now being done in an aqueous mixture of proteins and sugars. Cancer cells and zebrafish replaced round-bottom flasks." (Devaraj, 2018) This shift in thinking has led to a flurry of research by Bertozzi and others on how to modify desirable reactions from controlled environments so that they can also be performed in biological environments. For example, Bertozzi was awarded the Solvay Prize for demonstrating how

> to do click chemistry without copper, which is cytotoxic to cells. It is now possible to do some reactions . . . that are not harmful to the cell and that can

allow to develop further tagging any, why not, curing of the cells. ("Solvay Prize 2020, Carolyn Bertozzi Wins for Bioorthogonal Chemistry," 2020)

Scientific Impact

Interest in bioorthogonal chemistry has been exploding since Bertozzi and her colleagues introduced the phrase in 2003. Figure 5.4 shows the rapid growth in the number of publications that have the word "bioorthogonal" in their titles, abstracts, or author keywords since the term was introduced in 2003. Researchers have begun introducing many new bioorthogonal reactions,[2] and kits are now available to help nonexperts use bioorthogonal chemistry (Sletten and Bertozzi, 2011). It seems clear that the concept of bioorthogonal chemistry has rapidly evolved into a subfield in its own right.

Figure 5.4. Scientific Publications on Bioorthogonal Chemistry

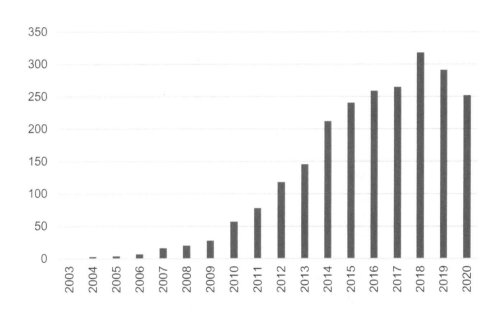

SOURCE: WOS core collection. Y-axis reflects the number of publications that have the word bioorthogonal in their titles, abstracts, or author keywords. Based on a similar figure by Sletten and Bertozzi, 2011.

Bertozzi has also continued to publish frequently, having more than 300 WOS publications. Figure 5.5 depicts Bertozzi's annual publication output from 1990 to 2019 (the most recent year for which complete data are available), and Table 5.2 provides statistics on her publications and citations during that time. Bertozzi's scientific publications have been cited more than 29,000 times by WOS-indexed publications. The journals in which Bertozzi's research has appeared most frequently are *Journal of the American Chemical Society* (51 articles), *Proceedings of the National Academy of Sciences of the United States of America* (43 articles), and *Angewandte*

---

[2] For recent examples, see An et al., 2018; Fang et al., 2018; Matsuo et al., 2018; and Yu et al., 2018.

Chemie International Edition (21 articles). The WOS categories into which Bertozzi's scientific publications are most frequently classified are biochemistry and molecular biology (100 articles); chemistry, multidisciplinary (89 articles); multidisciplinary sciences (57 articles); chemistry, organic (33 articles); and biochemical research methods (23 articles).

Figure 5.5. Bertozzi's Web of Science Publications by Year, 1990–2019

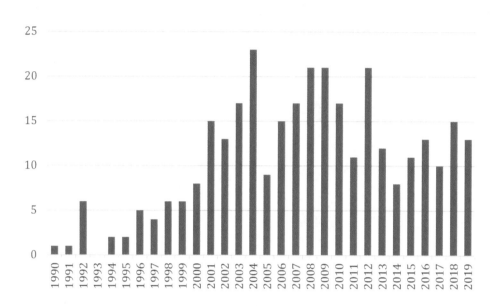

Table 5.2. Summary of Bertozzi's Publications

Total publications	324
Citations	29,498
Number of unique journals	90
WOS categories	35

SOURCE: WOS search conducted on March 26, 2020.

Bertozzi has received numerous awards for her scientific contributions. She is a member of the National Academy of Medicine, the National Academy of Sciences, the American Academy of Arts and Sciences, the German Academy of Sciences Leopoldina, and the National Academy of Inventors. Bertozzi has amassed many honors and awards in addition to the 2010 Lemelson-MIT Prize, including the Presidential Early Career Award in Science and Engineering, the Whistler Award, the Ernst Schering Prize, the Arthur C. Cope Award, the Heinrich Wieland Prize, a MacArthur Foundation Fellowship, the Solvay Prize, and dozens of other honors and awards. She has also received honorary doctorates from Brown University (in 2012), Freie University Berlin (in 2014), and Duke University (in 2014). Recognition has not been limited to her research. Bertozzi is widely regarded as an excellent mentor and teacher, and was awarded both the University of California, Berkeley Distinguished Teaching Award and the Donald Sterling Noyce Prize for Excellence in Undergraduate Teaching.

Over the course of her research, Bertozzi has co-authored with scholars from 26 distinct countries, as depicted in Figure 5.6. Her most frequent collaborators for scientific publications include David Rabuka (one of Bertozzi's graduate students from her time at University of California, Berkeley; 14 co-authored papers), Jeremy Baskin (another alumnus of Bertozzi's University of California, Berkeley lab; 13 co-authored papers), Steven Rosen (a biology professor whom Bertozzi worked for as a postdoctoral researcher at UCSF, and with whom she co-founded her first company; 12 co-authored papers), and Ellen Sletten (another alumni of Bertozzi's University of California, Berkeley lab; 12 co-authored papers).

Figure 5.6. Bertozzi's International Collaboration on Scientific Publications

NOTE: This networks graph was made using the bibliometrix package in R and the following settings: Network layout (star), clustering algorithm (none). Isolates have been removed and the thickness of the edges is weighted to reflect the number of co-authorships between the nodes. The graphs depict all co-authorship dyads, so ties between non-U.S. countries reflect publications in which at least one co-author from each country was also listed on a publication with Bertozzi.

Technological Impact

Table 5.3 summarizes the technological impact of Bertozzi's patent portfolio. As of January 1, 2020, Bertozzi held 64 patents and 23 published patent applications. Despite the nascent nature of Bertozzi's work, her patents already have 305 forward citations by 90 citing entities. Based on Bertozzi's continued work and the continued growth of this field of science, continued growth of these metrics seems likely.

Table 5.3. Summary of Bertozzi's Technological Impact

Total patents	87
Forward citations	305
Citing entities	90

Bertozzi's collective patent portfolio spans a variety of disciplines similar to Boyer and Cohen. Figure 5.7 provides the distribution of Bertozzi's patents by WIPO technology sector. Approximately 60 percent of Bertozzi's patents are classified under chemistry, and 36 percent are classified under instruments.

Figure 5.7. Distribution of Bertozzi's Patents by Technology Sector

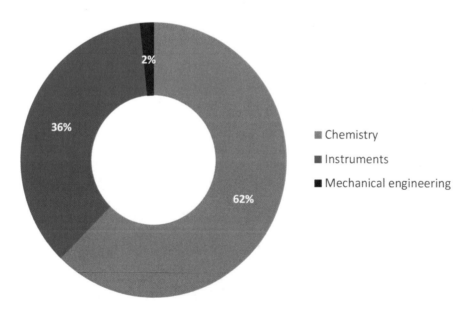

SOURCE: RAND analysis using patent data provided by IPVision, Inc.
NOTE: WIPO technology sector determined by IPC-WIPO sector crosswalk (Schmoch, 2008).

Like the other case studies, the technological impact of Bertozzi's invention can already be seen through the forward citations of future patent applications to Bertozzi's inventions. The organizations with the most forward citations are Forma Therapeutics (a clinical-stage biopharmaceutical company), Sofradim Production (a French medical equipment company), Glycomimetics (an oncology-focused biotechnology company), and Novo Nordisk (a major multinational pharmaceutical company). Similar to the other case studies, multiple research institutions are also building on Bertozzi's patents, including the Wisconsin Alumni Research Foundation, Auburn University, the Paul Scherrer Institut, the University of Arkansas, and the Children's Hospital & Research Center at Oakland. The separate appendix provides a summary of the top citing organizations to Bertozzi's patent portfolio.

Table 5.4 provides examples of new inventions among the top 25 citing organizations (i.e., entities with the highest number of forward citations) to Bertozzi's patent portfolio that would, in turn, lead to new discoveries and patents, each having ten or more forward citations themselves.

Table 5.4. Notable Inventions with Forward Citations to Bertozzi's Patent Portfolio

Citing Organization	Total Patents with Forward Citations	Examples of Notable Inventions	Year
Forma Therapeutics	24	3-aryl bicyclic [4,5,0] hydroxamic acids as histone deacetylase inhibitors	2018
Sofradim Production	23	Method of forming a medical device on biological tissue	2013
Glycomimetics	22	Compounds and methods for inhibiting selectin-mediated function	2006
		Heterobifunctional pan-selectin inhibitors	2010
Novo Nordisk	22	Intracellular formation of peptide conjugates	2010
		Factor VII: remodeling and glycoconjugation of Factor VII	2011
Covidien	14	Bioadhesive composition formed using click chemistry	2011
Zimmer	10	Methods of preparing hydrogel coatings	2012
Enzon	9	Single-chain antigen-binding proteins capable of glycosylation, production and uses thereof	2004
Life Technologies	8	Labeling and detection of nucleic acids	2012
Seattle Genetics	8	Methods and compositions for making antibodies and antibody derivatives with reduced core fucosylation	2012
Wisconsin Alumni Research Foundation	6	Chemical synthesis of reagents for peptide coupling	2005
Auburn University	6	Controlled and extended delivery of hyaluronic acid and comfort molecules via a contact lens platform	2013
Paul Scherrer Institut	6	Enzymatic conjugation of polypeptides	2016
3M Innovative Properties	5	Diacetylenic materials for sensing applications	2005
ratiopharm	5	Liquid formulation of G-CSF conjugate	2012

SOURCE: RAND analysis of data provided by IPVision, Inc. Notable inventions selected based on their own count of forward citations.

81

Economic and Social Impacts

In addition to being a highly regarded academic researcher, Bertozzi has developed her own approach to converting her team's scientific discoveries into new businesses. She has co-founded seven different companies, often in collaboration with her students or colleagues; most have been founded since 2014. Once each company has been founded, Bertozzi has generally settled into an advisory role rather than a management one. Her typical role is scientific adviser or chair of scientific advisory board, which allows her to focus on continuing to advance scientific research.

Because most of these businesses are only a handful of years old, their ability to viably convert scientific research into commercial products is unknown. Although much uncertainty remains about the future economic impact of these organizations, this section discusses each organization's activities and describes their potential impacts.

Thios Pharmaceuticals

The first company Bertozzi co-founded was Thios Pharmaceuticals, which she co-founded with Steve Rosen, Stefan Hemmerich (another former postdoctoral researcher of Rosen's who was then at Roche Biosciences), and Ted Yednock (then at Elan Pharmaceuticals). During her postdoctoral research, Rosen and Bertozzi discussed the idea of forming a company to pursue drug development based on their research on sulfation. *Sulfation* is "the process in which sulfate recovered by cells from the bloodstream is attached to target molecules, such as glycoproteins." (Jacobson, 2002). This focus was reflected in the company's name; *thios* is the Greek word for sulfur. Bertozzi ended up leaving Rosen's lab much sooner than she had originally anticipated to join the faculty at University of California, Berkeley, but after she settled in at the university, the two revisited the idea, founding Thios in 2001 (Bertozzi, 2003).

In 2002, the company received $15 million in support from the life science venture capital firms HealthCare Ventures and Skyline Ventures ("Thios Pharmaceuticals Inc.," undated). By 2004, 24 people were involved in the company (McCarthy, 2004). Bruce A. Hironaka, then the president and CEO of Thios Pharmaceuticals, noted that "[o]ur mission is to discover, develop, and commercialize innovative therapies targeting biological sulfation to treat serious unmet medical needs." The company was pursuing two different paths to create these therapies. One effort focused on using sulfated molecules as small-molecule drugs, while the other focused on using sulfated molecules as targets for an antibody drug. In an ideal world, they hoped to "echo the commercial success of kinase inhibitors, such as the cancer therapy, Gleevec. Gleevec was FDA approved in 2001 for treatment of chronic myeloid leukemia and in 2003 generated sales of over $1.1 billion" (McCarthy, 2004).

As is often the case for new start-ups, Thios Pharmaceuticals did not ultimately succeed in developing new medical treatments; the company appears to have become inoperative. Like DeSimone's example in Chapter 2, Bertozzi's first foray into founding a company did not result in a successfully established business, but the scientific impact of the work remains. Thios produced and influenced multiple publications since its closure (Arata-Kawai et al., 2011; Yang et al., 2006) that may yet impact future work; much of Bertozzi's own bioorthogonal research builds upon reactions discovered many decades prior that had few obvious applications at the time.

Redwood Bioscience

Bertozzi's second foray into the commercialization of her research met with greater success. Bertozzi and her graduate student, David Rabuka, founded Redwood Bioscience with the goal of commercializing a technology they called SMARTag, a site-specific protein modification tool. The technologies developed by Redwood "enable the generation of homogenous bioconjugates engineered to improve performance and ease of manufacturing." This "enables the identification of superior drugs from libraries of differentially designed conjugates," and "the generation of homogenous antibody-drug conjugates engineered to enhance potency, safety and stability" (Garde, 2014).

The budding SMARTag technology attracted the support of Catalent Pharma Solutions, a company that develops drug delivery technologies to help develop pharmaceuticals, biologics, and consumer health products. Redwood's SMARTag technology complemented Catalent's existing gene-insertion technology for manufacturing pharmaceuticals. In April 2013, Catalent acquired an exclusive license to market the SMARTag technology and collaborated with Redwood to support further development of the technology. Catalent's interest in the technology grew steadily; in March 2014, Catalent increased its minority stake in Redwood, and in October 2014, Catalent acquired the remaining stake in Redwood. Financial details of the acquisition were not disclosed. Following the acquisition, Rabuka spent several years as global head of research and development at Catalent and is now the founder and CEO of Acrigen Biosciences. Bertozzi remains on the biologics advisory board and Redwood scientific advisory board at Catalent.

Catalent has acquired many biotech companies since its July 2014 IPO. It now has more than 1,000 products in development at any time, launches more than 180 new products annually, employs 14,000 people around the world, and generated more than $3 billion of revenue in 2020 (Catalent, undated-a, undated-b). Separating out the portion of those metrics attributable to Bertozzi's inventions is not feasible based on publicly available information. The ultimate economic impact of the technology developed by Bertozzi and Rabuka remains to be seen, but appears promising. The first Phase 1 clinical trial that applies the technology is currently underway (Triphase Research and Development III Corporation, 2020); it targets heavily pretreated patients with relapsed/refractory B-cell non-Hodgkin's lymphoma. Non-Hodgkin's lymphoma is one of the most common types of cancer in the United States and is often fatal; the disease kills five out of 100,000 people in the United States each year (Centers for Disease Control and Prevention, undated-b). Animal testing, along with more-recent early low-level dose testing in humans, suggest that the developed drug is safe and effective; no significant adverse events have occurred and early signs of efficacy were observed (Drake, 2020). Further test results will be needed to precisely estimate the potential economic impact of this drug, but the treatment's potential—and the potential of SMARTag technology to help develop additional treatments for additional diseases—seems promising.

Enable Biosciences

In 2015, Bertozzi co-founded a third company, Enable Bioscience, along with Jason Tsai, Peter Robinson, and David Seftel; Tsai and Robinson are both graduate students in Bertozzi's lab. Enable Biosciences focuses on antibody-based disease diagnostics. The team developed the ADAP technology, which couples specific DNA tags with antigens that are used to detect the presence of antibodies in blood, saliva, or other bodily fluids. The potential economic value of these tests comes from the earlier identification of diseases, and any associated avoidance of complications or improvement in treatment outcomes.

Enable Biosciences is a small start-up in San Francisco with a pipeline of several ADAP tests. Of these tests, one—a test for predicting type 1 diabetes—is now available for clinical diagnostics. Other tests under development (and available for research use) use ADAP to test for food allergies, celiac disease, Lyme disease, Zika and dengue fevers, and severe acute respiratory syndrome coronavirus 2 (SARS-CoV-2, otherwise known as coronavirus disease 2019 [COVID-19]; Enable Biosciences, undated-a, undated-b). Enable Biosciences' SARS-CoV-2 test is listed by the FDA and was listed as under review for Emergency Use Authorization as of October 29, 2020 (Enable Biosciences, undated-a).

To illustrate the potential value of these tests, we discuss the potential economic value of the type 1 diabetes test in further detail. Estimates of the number of new type 1 diabetes cases that are diagnosed each year vary; the 2020 National Diabetes Statistics Report states that just more than 18,000 children and adolescents younger than 20 years old are diagnosed with type 1 diabetes each year (Centers for Disease Control and Prevention, 2020a), while Rogers et al. (2017) estimate that there are 27,000 new cases each year among children and adolescents younger than 20 years old, as well as another 37,000 new cases each year among adults 20 to 64 years old. Patients are generally unaware that they have type 1 diabetes until they begin to feel ill. Symptoms often appear similar to flu or malaise (JDRF, undated), but, partly because of the unanticipated nature of onset, 30 to 46 percent of children suffer from diabetic ketoacidosis at the time of their initial diagnosis (Duca et al., 2017). Diabetic ketoacidosis is a life-threatening condition caused by the lack of insulin that requires costly hospital-based care.

There is currently no cure for type 1 diabetes or behavioral change that can prevent it, but carefully managed insulin therapy enables patients to avoid serious side effects. Fortunately, there have been recent advances in the ability to predict type 1 diabetes. As an autoimmune disease, the presence of any of four particular autoantibodies is known to be a risk factor for type 1 diabetes, and the presence of any two or more of these autoantibodies generally causes type 1 diabetes to occur (Ziegler et al., 2013). These autoantibodies can be present well in advance of disease symptoms; they can be present at as early as six months of age (Ziegler et al., 2012; Parikka et al., 2012; Krischer et al., 2015), and "by three years of age, the majority of patients in whom clinical type 1 diabetes ultimately will develop during childhood will be islet autoantibody positive." (Bonifacio, 2015) This has led experts to recommend that all children be tested for these autoantibodies at three years of age; Enable Bioscience has now developed such a test and has received regulatory approval for its use in clinical diagnostics.

The ultimate benefits of this test, as well as the benefits of Enable Bioscience's pipeline of other tests, remain to be seen. Bonifacio (2015) explains that

> [b]enefits range from learning about the disease process to preventing complications such as diabetic ketoacidosis at the diagnosis of diabetes, or even the prevention of diabetes entirely. Costs are usually pinned to the anxiety that may be caused in families who have a positive biomarker result. The long-term cost/benefit ratio of estimating the risk of type 1 diabetes risk is still under investigation.

Under presently available technology, widespread use of the Enable Bioscience test could help thousands of children avoid diabetic ketoacidosis every year; costs would include the production, distribution, and application of the tests. If such testing became standard and the test could be cheaply mass-produced, then the technology could be well positioned to provide meaningful economic benefits. Should researchers eventually develop a method for delaying or preventing the onset of type 1 diabetes, the economic value of a predictive test would be enhanced significantly.

Palleon Pharmaceuticals

In 2015, Palleon Pharmaceuticals was founded by Jim Broderick (a biotech entrepreneur who had successfully co-founded and led several prior biotech companies), Bertozzi, and Paul Crocker (professor of glycoimmunology and head of the Division of Cell Signaling and Immunology at the University of Dundee in Scotland). Palleon is using the latest advancements in glycoscience to develop new methods for treating cancer. Palleon currently lists seven different drug programs in their pipeline of therapeutic candidates; all of these drugs are currently in preclinical testing.

Palleon's pipeline builds upon the scientific discoveries and advancements surrounding the cell surface glycan patterns on cancer cells, and the role of those surfaces in preventing the body's natural immune reaction from destroying tumors. The activating receptor complex on an immune cell would bind to the activating ligand on the cancer cell, triggering the body's natural immune response. However, a series of receptors known as *Siglecs* turn off that response whenever they bind to any one of many sialic acid–containing glycans that cover the surface of the cancer cell. Targeting this "off switch" has been difficult because there are many different types of Siglecs and many different types of sialic acid–containing glycans, resulting in a large array of combinations that can trigger the switch and prevent the body's immune system from attacking the cancer cell.

Several of Palleon's under-development drugs seek to address this challenge by removing the terminal sialic acids from cell surface glycans, regardless of the precise shape of the glycan. Without the sialic acids, the Siglecs no longer bind, and the switch is no longer triggered, allowing the body's natural immune reaction to proceed. Animal models and nonanimal laboratory tests of Palleon's lead drug program, Sialidase-Fc, have been promising, and the drug is expected to enter clinical testing in 2021.

Even if all goes well, many steps remain before this technology is ready for widespread use by the general public. As with most start-up biotechnology firms, it will take many years to convert the scientific advancements into monetizable benefits. Nevertheless, investor support offers a market-based signal of potential returns on the technology. In September 2020, Palleon announced that it had raised $100 million in Series B financing—a vote of confidence that the potential returns to these inventions would exceed $100 million. The supporting investors included a wide variety of prominent venture capital firms, including SR One, Pfizer Ventures, Vertex Ventures HC, Takeda Ventures, AbbVie Ventures, and Surveyor Capital (Sarkis, 2020).

InterVenn Biosciences

InterVenn Biosciences was co-founded in 2017 by Aldo Carrascoso (CEO of InterVenn and founder of several collaboration and technology companies), Lieza Danan (then the head of mass spectrometry at Stemcentrx and previously a Ph.D. student and postdoctoral researcher at University of California, Davis, with focuses on bioorthogonal chemistry and mass spectrometry), Carlito Lebrilla (distinguished professor of chemistry at University of California, Davis, where his laboratory focuses on mass spectrometry), and Carolyn Bertozzi.

InterVenn combines artificial intelligence with mass spectrometry to create technology platforms that support the diagnosis and monitoring of different diseases, as well as prediction of the effectiveness of potential therapies. The company has six products in its research pipeline, with a focus on cancer. As with other new biotechnology companies, InterVenn's products are not yet commercially available. Their most advanced product, a blood test that enables earlier detection of ovarian cancer, is undergoing clinical validation.

Ovarian cancer has significantly higher survival rates if treated before it spreads to additional parts of the body, but only 20 percent of ovarian cancers are detected in earlier stages because the disease is typically asymptomatic in early stages and an effective screening test is not available.[3] Ovarian cancer is the fifth-most common cause of cancer deaths among women in the

---

[3] An ultrasound can be used to identify whether a mass is present but cannot determine whether the mass is cancerous or benign. A blood test that measures amounts of a protein known as CA-125 can help track the size of tumors during treatment, but "has not been found to be as useful as a screening test for ovarian cancer" (American Cancer Society, 2020a). Survival rate statistics available at American Cancer Society, 2020b.

United States; one in 78 women experience malignant ovarian cancer, and the disease is the cause of death for one in 108 women (American Cancer Society, 2021). Full results of the clinical trial are expected in April 2021, but initial results suggest InterVenn's blood test was "consistently able to differentiate malignant from benign tumors," according to Carrascoso (Vuturo, 2019). Although the potential impact of the drug depends on many factors, including its accuracy rate and the extent to which the test becomes standard practice, earlier detection of ovarian cancer could save thousands of lives every year.[4]

A collection of biotech investors (Genoa Ventures, True Ventures, Amplify Partners, Boost VC, and Prado SV) have provided $9.4 million to support the development of this ovarian cancer blood test (Carrascoso, 2018). InterVenn remains a small company with less than 25 employees, but is growing rapidly, with ten currently posted job openings.

OliLux Biosciences

In 2019, Olilux Biosciences was founded by Mireille Kamariza (CEO of Olilux and a student from Bertozzi's lab), Bertozzi, and Manu Prakash (professor of bioengineering at Stanford, where he focuses on low-cost innovations that make science and medicine accessible to developing countries). This new company is focused on developing a rapid, low-cost diagnostic test for tuberculosis. The test is based on research led by Kamariza and Bertozzi (Kamariza et al., 2018); they created a dye that fluoresces within one hour if it is incorporated with the live mycobacteria that causes tuberculosis. The dye, which is designed to be mixed with sputum (mucus coughed up from the trachea or lungs), does not fluoresce if tuberculosis bacteria are not present; it does not require additional processing steps.

Although other tests for tuberculosis exist, they are generally more expensive, more complex, or less effective. In addition, because the dye only reacts to live mycobacteria, it can be used as an affordable method not only to find if a person has tuberculosis, but also to test whether treatments are working or whether the patient is contagious. A vaccine is available for tuberculosis, and a variety of drugs are available to cure cases of the disease, but the drugs must be taken for six to nine months. Some strains of tuberculosis are becoming resistant to certain drugs because patients stop taking the drugs too soon (Centers for Disease Control and Prevention, 2016).

The number of tuberculosis cases in the United States has declined considerably, with less than three per 100,000 people becoming infected each year (Centers for Disease Control and

---

[4] The American Cancer Society estimates that 21,750 women will be diagnosed with ovarian cancer in 2020, although this rate has been falling slowly over time. Eighty percent of cases (17,400) are not detected in earlier stages. Survival rates while the cancer is localized vary from 92 to 98 percent, depending on the type of ovarian cancer; if the disease has spread, survival rates vary from 30 to 94 percent, depending on the type of ovarian cancer and the extent to which the disease has spread to other parts of the body (American Cancer Society, 2021).

Prevention, 2020b). However, the disease continues to ravage developing countries,[5] and its worldwide impact on mortality is severe. Tuberculosis "is one of the top ten causes of death and the leading cause from a single infectious agent" (World Health Organization, 2020). The World Health Organization estimates that in 2019, 10 million people fell ill with tuberculosis[6] and 1.4 million died (World Health Organization, 2020). International efforts to end tuberculosis are underway. Although the impact that OliLux's diagnostic test might have is not yet clear, it could provide a valuable tool in the global fight against tuberculosis.

Lycia Therapeutics

Lycia Therapeutics is Bertozzi's newest company; very little information about the company has been publicly released. The company is a collaboration between Versant Ventures and Bertozzi; it was founded in 2019 by Versant's Inception Therapeutics. In March 2020, Aetna Wun Trombley stepped down as president and chief operating officer at NGM Bio to become CEO at what was then "an undisclosed privately held company" ("Aetna Wun Trombley . . .," 2020). In June 2020, the existence of Lycia Therapeutics was publicly announced and Versant Ventures pledged $50 million in support of the new company, indicating Versant's high expectations of the Lycia's potential.

Like much of Bertozzi's work, the firm appears to be focused on exploiting interactions with cell surface proteins. Lycia has publicly announced that they are focused on developing "lysosomal targeting chimeras, or LYTACs, as therapeutics for a broad set of currently intractable cell surface targets" (Edelson, 2020). One mechanism of focus is "a cation-independent receptor called CI-M6PR [that] could be exploited to capture and drag extracellular proteins into cells, trafficking them to the lysosome for destruction" (Edelson, 2020), on which Bertozzi has recently published several papers (Ahn et al., 2020; Banik et al., 2019). It has been suggested that Lycia has now advanced beyond the information described in Bertozzi's publication (Tong, 2020). At this point, further details on the company and its planned platforms or products are not publicly available; therefore, we cannot assess its potential impacts.

Involvement in Other Organizations

In addition to co-founding the companies described above, Bertozzi has also contributed her time and expertise to a variety of other organizations. She has served as an adviser to many major companies, including the research advisory board at GlaxoSmithKline and the Board of directors at Eli Lilly & Co. She also serves as a scientific adviser at Glympse Bio, which was co-

---

[5] "Eight countries accounted for two thirds of the new [tuberculosis] cases: India, Indonesia, China, Philippines, Pakistan, Nigeria, Bangladesh and South Africa" (World Health Organization, 2020).

[6] Only 5 to 15 percent of those who become infected with the mycobacteria that cause tuberculosis become ill at some point in their life. Individuals with compromised immune systems, such as those with HIV, are at higher risk of becoming ill (World Health Organization, 2020).

founded by another Lemelson-MIT Prize Winner, Sangeeta Bhatia. Bertozzi has also worked with the Grace Science Foundation, which supports research on rare diseases.

Conclusion

Bertozzi's work has already created significant scientific advancements, and those advancements are actively being converted into new technologies. However, the development of such new biomedical technologies takes a significant amount of time, and although these technologies have great potential, the economic impacts of Bertozzi's work are yet to be fully realized.

Bertozzi represents a different style of inventor: She has generally eschewed the CEO role in favor of positions that allow her to focus on advancing science while others handle day-to-day business operations. Bertozzi's approach focuses on identifying gaps in underlying scientific knowledge, investigating those gaps, and then developing new technologies based on the advancements in scientific knowledge. Furthermore, her work has the potential for widespread applicability due to its generalizability. Her laboratory focuses on creating generalizable platform technologies that exploit scientific advancements with diverse uses, rather than directly focusing on a solution specific to a single issue. Therefore, Bertozzi's overall approach to invention might be best described as enabling future inventors of specific biomedical treatments by discovering and inventing the tools they will use to create those biomedical inventions.

Bertozzi also shows that inventors can create lasting impacts through channels beyond the direct impacts of their inventions. Bertozzi is widely regarded as an excellent mentor to students of all ages, including postdoctoral fellows, graduate students, undergraduate students, and K–12 students. Many of Bertozzi's companies and scientific publications are collaborations with students whom she has mentored, and many of her students have gone on to become highly regarded chemists in their own right.

6. Conclusion

In this report, we examine the impact of the inventions of a particularly high-achieving group of inventors, the winners of the Lemelson-MIT Prize, both in aggregate and through three case studies. These inventions changed the world by solving very difficult problems, enabling easier solutions, and improving how society lives, learns, heals, communicates, and plays. In some cases, these inventions were revolutionary ideas that spawned entirely new industries. These impacts are reflected in the citation of the inventors' works in scientific publications, new product patents, licensing, or through open-source technologies. As of March 2020, the 26 prize winners had published more than 3,700 articles that had accumulated more than 334,000 citations in the WOS citation index. As of January 1, 2020, their 3,871 patents had been cited as prior art by more than 40,000 subsequent patents.

We identified more than 180 organizations affiliated with the 26 Lemelson-MIT Prize winners and their inventions. As described in Chapter 2, two companies founded by Lemelson-MIT Prize winners were among the first biotechnology firms to achieve a market capitalization exceeding $100 billion. Several other publicly traded firms founded by Lemelson-MIT Prize winners had market valuations between $100 million and $50 billion as of 2020. As of 2019, 35 other companies founded by prize winners had been acquired or had merged with other business entities in market transactions valued at approximately $7.5 billion (in 2019 dollars), with 22 more involved in acquisitions for which our research team was not able to find publicly available financial details. Of those companies that remained independent entities as of 2019, a small number grew into large profitable companies, while the vast majority remained successful smaller companies or were start-ups attempting to commercialize new ideas or technologies. Some companies did not succeed. Those independent companies that report financial data collectively employ approximately 40,000 people and generate total annual revenues exceeding $54 billion.

The combined technologies, resources, and capabilities of these organizations have achieved significant advances in several areas, including pharmaceutical manufacturing, drug delivery systems, pattern recognition based computer technology, mobility devices, and energy storage systems. The impacts of the inventions presented in this report may not reflect the impacts of an average invention, as the actual or anticipated impact of the inventions was integral to the selection of the winners of the Lemelson-MIT Prize. However, there have been and will continue to be many highly successful and impactful inventors beyond this small group.

Chapters 3 through 5 presented case studies on the Lemelson-MIT Prize winners from three years: 1996 (Herbert Boyer and Stanley Cohen), 2003 (Leroy Hood), and 2010 (Carolyn Bertozzi). These prize winners' inventions helped give rise to the modern biotechnology industry and advanced it in new and unforeseen ways. Their inventions created entirely new fields of study and revolutionized existing techniques including genetic engineering, genomics and proteomics, and bioorthogonal chemistry. In 2020, the U.S. biotechnology industry had approximately 2,200 total firms, 288,000 employees, and $108 billion in annual revenues (IBISWorld, 2020). The first biotechnology companies would have enormous successes, including some of the largest public offerings in U.S. history, as well as considerable failures. Their impacts paved the way for many new discoveries, products, and future entrepreneurs. Several themes arose from these case studies, as well as from reviews of the other inventors (as described in the separate appendix).

First, the impacts of many of the inventions go beyond the direct effect of the invention. These inventions are often platforms or proofs of concept that enable or inspire entirely new fields of study or industries. All three case studies examined in this report reflect cases where the basic research techniques, scientific instruments, or new medical tools spawned a revolution in the types of approaches that could be taken to address significant health challenges. Even though these case studies reflect the development of new ideas, they still built upon the work that came before them, which emphasizes the difficulty in attributing the impacts of the final outcome to any single entity or individual.

Second, successful inventors help solve major challenges facing society. The inventions described in Chapters 3 through 5 continue to help address a wide variety of diseases. The role of invention in addressing global health challenges continues to play out today, as inventors help tackle challenges created by the current COVID-19 pandemic. Several of the companies founded by Lemelson-MIT Prize winners have efforts underway to address challenges related to the pandemic. For example, Enable Biosciences (one of the companies co-founded by Bertozzi) has developed a blood test with a 94.1 percent specificity for sensing SARS-CoV-2 neutralizing antibodies within the first 15 days of onset (Enable Biosciences, undated-a). It has partnered with the California Department of Public Health; the Cerus Corporation; the Vaccine Development Research Laboratory at the University of California, Irvine; the Vitalant Research Institute; and the California National Primate Research Center to work on optimizing convalescent plasma therapy for COVID-19 patients (Lee, 2020). InterVenn Biosciences (another company co-founded by Bertozzi), announced in October 2020 that it had "identified marked differences in the glycoproteomic profile of patients who became seriously ill with COVID-19, as compared to individuals who had also been infected with the SARS-CoV-2 virus but experienced no or minimal symptoms" (Vuturo, 2020). The role of inventions in addressing public health crises is not limited to medical inventions. Another Lemelson-MIT Prize winner, Ramesh Raskar, founded the PathCheck Foundation, which seeks to use technology to support public health initiatives, such as contact tracing, while protecting user privacy and using open-source methods (Lee and Grant, 2020; PathCheck Foundation, undated).

Third, maximizing the impact of an invention requires more than creating an invention alone. Addressing challenges in applying and manufacturing an invention, as well as navigating regulatory, financial, and business-management hurdles, requires a talented team and (in some cases) a bit of luck. For example, Fogarty struggled to find a company willing to manufacture his balloon catheter, just as Hood struggled to find manufacturers for his early scientific instruments and Boyer struggled to conduct industrial trials of his new rDNA technology. Cohen and Boyer had to navigate legal challenges on the patentability of genetically modified organisms, as well as significant backlash from both the general public and the academic and scientific community. In some cases, a perfectly effective invention may not translate into a viable business model, such as DeSimone's environmentally friendly dry cleaning solvents. More than a dozen of the companies founded by Lemelson-MIT Prize winners eventually ceased to operate, often because of external factors (e.g., changes in market demand, the price of competing products) instead of a failure of the invention to perform as desired. It is also common for companies to be founded prior to the completion of any final consumer product, as is the case with many of Bertozzi's companies; this is one reason entrepreneurship can be risky. In many cases, inventors partnered with others to establish new organizations to tackle these challenges, again highlighting the difficulty in attributing the benefits of invention to the work of any single individual.

The scientific, technical, economic, and social benefits provided by inventions, along with the challenges faced by inventors, highlight the importance of inventorship. Evidence that a lack of exposure to role models and support networks is limiting the diversity of inventors emphasizes the importance of programs that support inventorship (Bell et al., 2019). Supporting the next generation of inventors is vital because invention is an ongoing social investment—not a largely depleted checklist of opportunities. Future work should be focused on shaping such efforts to address inequality and assessing their effectiveness.

Several notable efforts are already underway. For example, the Study of Underrepresented Classes Chasing Engineering and Science Success (SUCCESS) Act of 2018

> requires the Director of the [USPTO], in consultation with the Small Business
> Administration (SBA), to identify publicly available data on the number of
> patents annually applied for and obtained by women, minorities, and veterans
> and the benefits of increasing the number of patents applied for and obtained by
> women, minorities, and veterans and the small businesses they own. (Iancu and
> Peter, 2019)

The resulting congressionally mandated study (Iancu and Peter, 2019) includes potentially impactful recommendations for both the USPTO and congressional legislators, such as expanding educational outreach programs and workforce training programs. It also includes recommendations to enhance the data collection and data-sharing capabilities needed to track the effectiveness of efforts to increase diversity among inventors. The literature suggests that engaging with potential inventors at a young age may be particularly important for future outcomes (Chetty et al., 2011; Bell et al., 2019). Based on our findings that even successful inventors commonly struggle to bring new inventions to market, policymakers might consider supporting interventions that help new entrepreneurs, particularly those from historically disadvantaged groups, navigate the challenges of establishing a new business. Efforts such as these could play a role in broader efforts to diversify participation in science, technology, engineering, and mathematics fields. Efforts that increase the diversity of inventors are important because the benefits of invention will be greatest when all members of society are encouraged to contribute their creative approaches for addressing tomorrow's challenges.

Bibliography

Accelerator Life Science Partners, "About," webpage, undated. As of November 10, 2020:
http://www.acceleratorlsp.com/about

"Aetna Wun Trombley, Ph.D. to Step Down as President and Chief Operating Officer of NGM Bio to Assume Chief Executive Officer Role at Undisclosed Privately Held Company," *Yahoo News!* March 9, 2020. As of January 25, 2021:
https://finance.yahoo.com/news/aetna-wun-trombley-ph-d-203010107.html

Agarwal, Paresh, Brendan J. Beahm, Peyton Shieh, and Carolyn R. Bertozzi, "Systemic Fluorescence Imaging of Zebrafish Glycans with Bioorthogonal Chemistry," *Angewandte Chemie International Edition*, Vol. 54, No. 39, 2015, pp. 11504–11510.

Agilent Technologies, *2019 Agilent Technologies Inc. Proxy Statement*, 2019. As of January 29, 2021:
https://s24.q4cdn.com/305549747/files/doc_financial/annual/2018/Proxy-Web-DFIN.pdf

Ahn, Green, Steven Banik, Caitlyn L. Miller, Nicholas Riley, Jennifer R. Cochran, and Carolyn Bertozzi, "Lysosome Targeting Chimeras (Lytacs) that Engage a Liver-Specific Asialoglycoprotein Receptor for Targeted Protein Degradation," *ChemRxiv*, 2020.

Albert, M. B., D. Avery, F. Narin, and P. McAllister, "Direct Validation of Citation Counts as Indicators of Industrially Important Patents," *Research Policy*, Vol. 20, No. 3, 1991, pp. 251–259.

American Cancer Society, "Can Ovarian Cancer Be Found Early?" webpage, 2020a. As of January 25, 2021:
https://www.cancer.org/cancer/ovarian-cancer/detection-diagnosis-staging/detection.html

———, "Survival Rates for Ovarian Cancer," webpage, 2020b. As of January 25, 2021:
https://www.cancer.org/cancer/ovarian-cancer/detection-diagnosis-staging/survival-rates.html

———, "Key Statistics for Ovarian Cancer," webpage, 2021. As of January 28, 2021:
https://www.cancer.org/cancer/ovarian-cancer/about/key-statistics.html

Amgen, "Amgen History," webpage, undated. As of November 10, 2020:
https://www.amgen.com/about/amgen-history

———, Form 10-K Annual Reports, 1995–2019. Retrieved from U.S. Securities and Exchange Commission, Electronic Data Gathering, Analysis, and Retrieval System (EDGAR).

———, *2019 Letter to Shareholders*, 2020. As of January 13, 2021:
https://investors.amgen.com/static-files/d45cb739-9637-4e13-856c-e2fc29571032

An, Peng, Tracey M. Lewandowski, Tuğçe G. Erbay, Peng Liu, and Qing Lin, "Sterically Shielded, Stabilized Nitrile Imine for Rapid Bioorthogonal Protein Labeling in Live Cells," *Journal of the American Chemical Society*, Vol. 140, No. 14, 2018, pp. 4860–4868.

Applied Biosystems, "Advancing Science for Twenty-Five Years: Applied Biosystems Timeline," web page, undated. As of February 1, 2021:
https://web.archive.org/web/20070927044911/http:/marketing.appliedbiosystems.com/mk/get/25YRSEMS_HERRITAGE_TIMELINE

"Applied Biosystems, Invitrogen Complete $6.7 Billion Merger," *San Francisco Business Times*, November 21, 2008. As of November 10, 2020:
https://www.bizjournals.com/sanfrancisco/stories/2008/11/17/daily103.html

Arata-Kawai, Hanayo, Mark S. Singer, Annette Bistrup, Annemieke van Zante, Yang-Qing Wang, Yuki Ito, Xingfeng Bao, Stefan Hemmerich, Minoru Fukuda, and Steven D. Rosen, "Functional Contributions of N- and O-Glycans to L-Selectin Ligands in Murine and Human Lymphoid Organs," *American Journal of Pathology*, Vol. 178, No. 1, 2011, pp. 423–433.

Arizona State University, "Sugar-Coated World: Probing the Mysteries of Glycan Structure and Function," *ScienceDaily*, December 18, 2017. As of January 28, 2021:
https://www.sciencedaily.com/releases/2017/12/171218120356.htm

Banik, Steven, Kayvon Pedram, Simon Wisnovsky, Nicholas Riley, and Carolyn Bertozzi, "Lysosome Targeting Chimeras (Lytacs) for the Degradation of Secreted and Membrane Proteins," *ChemRxiv*, 2019.

Barany, Francis, "Genetic Disease Detection and DNA Amplification Using Cloned Thermostable Ligase," *Proceedings of the National Academy of Sciences of the United States of America*, Vol. 88, No. 1, 1991, pp. 189–193.

Baskin, Jeremy M., Karen W. Dehnert, Scott T. Laughlin, Sharon L. Amacher, and Carolyn R. Bertozzi, "Visualizing Enveloping Layer Glycans During Zebrafish Early Embryogenesis," *Proceedings of the National Academy of Sciences of the United States of America*, Vol. 107, No. 23, 2010, pp. 10360–10365.

Battelle Memorial Institute, *The Impact of Genomics on the U.S. Economy*, Columbus, Ohio, 2013. As of November 10, 2020:
http://unitedformedicalresearch.org/wp-content/uploads/2013/06/The-Impact-of-Genomics-on-the-US-Economy.pdf

Beaulieu, Luc, "How Many Citations Are Actually a Lot of Citations?" *Ruminating . . .*, blog post, November 20, 2015. As of January 13, 2021: https://lucbeaulieu.com/2015/11/19/how-many-citations-are-actually-a-lot-of-citations

Bell, Alex, Raj Chetty, Xavier Jaravel, Neviana Petkova, and John Van Reenen, "Who Becomes an Inventor in America? The Importance of Exposure to Innovation," *Quarterly Journal of Economics*, Vol. 134, No. 2, 2019, pp. 647–713.

Bell, Julie, "Celera Genomics Completes Acquisition of Calif. Company; Paracel Products Used to Search Gene Databases; Genome Research," *Baltimore Sun*, June 10, 2000.

Bennett, J. Claude, Leroy E. Hood, William J. Dreyer, and Michael Potter, "Evidence for Amino Acid Sequence Differences Among Proteins Resembling the L-Chain Subunits of Immunoglobulins," *Journal of Molecular Biology*, Vol. 12, No. 1, 1965, pp. 81–87.

Berg, Paul, transcript of an interview conducted by Rae Goodell at the Massachusetts Institute of Technology, Cambridge, Mass., April 17, 1978.

Berkrot, Bill, and Susan Kelly, "Thermo Fisher to Buy Life Tech for $13.6 Billion," Reuters, April 15, 2013. As of December 21, 2020: https://www.reuters.com/article/us-lifetechnologies-thermofisher/thermo-fisher-to-buy-life-tech-for-13-6-billion-idUSBRE93D0A620130415

Bertozzi, Carolyn R., transcript of an interview conducted by Andrea R. Mastrejuan at the University of California, Berkeley, August 17–18, 2003.

———, "A Decade of Bioorthogonal Chemistry," *Accounts of Chemical Research*, Vol. 44, No. 9, 2011, pp. 651–653.

Bessen, Jeff, "GMOs Could Save Your Life—They Might Have Already," *New Republic*, July 28, 2016.

BioSpectrum, "25 Years of Applied Biosystems' Model 470A Protein Sequencer," press release, October 5, 2007. As of October 20, 2020: https://www.biospectrumindia.com/news/73/2843/25-years-of-applied-biosystems-model-470a-protein-sequencer.html

Blind, Knut, Katrin Cremers, and Elisabeth Mueller, "The Influence of Strategic Patenting on Companies' Patent Portfolios," *Research Policy*, Vol. 38, No. 2, 2009, pp. 428–436.

Bolivar, Francisco, Raymond L. Rodriguez, Patricia J. Greene, Mary C. Betlach, Herbert L. Heyneker, Herbert W. Boyer, Jorge H. Crosa, and Stanley Falkow, "Construction and Characterization of New Cloning Vehicles. II. A Multipurpose Cloning System," *Gene*, Vol. 2, No. 2, 1977, pp. 95–113.

Bonifacio, Ezio, "Predicting Type 1 Diabetes Using Biomarkers," *Diabetes Care*, Vol. 38, No. 6, 2015, pp. 989–996.

Bornmann, Lutz, "Scientific Peer Review," *Annual Review of Information Science and Technology*, Vol. 45, No. 1, 2011, pp. 197–245.

Brunt, Liam, Josh Lerner, and Tom Nicholas, "Inducement Prizes and Innovation," *Journal of Industrial Economics*, Vol. 60, No. 4, 2012, pp. 657–696.

BuiltWith, "reCAPTCHA Usage Statistics," webpage, undated. As of January 28, 2021:
https://trends.builtwith.com/widgets/reCAPTCHA

Cagan, Jonathan, and Craig M. Vogel, *Creating Breakthrough Products: Innovation from Product Planning to Program Approval*, Upper Saddle River, N.J.: Prentice-Hall, 2002.

Carrascoso, Aldo, "InterVenn BioSciences Raises $9.4 Million to Help Physicians Detect Early Ovarian Cancer with a Minimally Invasive Test," InterVenn, press release, December 12, 2018. As of January 25, 2021:
https://intervenn.com/blog/intervenn-biosciences-raises-9-4-million-to-help-physicians-detect-early-ovarian-cancer-with-a-minimally-invasive-test/

Catalent, homepage, undated-a. As of January 25, 2021:
https://www.catalent.com/about-us/overview

———, "Investor Relations," webpage, undated-b. As of January 25, 2021:
https://investor.catalent.com/investor-center-home/default.aspx

Centers for Disease Control and Prevention, "Diabetic Diagnosis," webpage, undated-a. As of December 31, 2020:
https://gis.cdc.gov/grasp/diabetes/DiabetesAtlas.html#

———, "United States Cancer Statistics: Data Visualizations, Leading Cancer Causes and Deaths, All Races/Ethnicities, Male and Female, 2017," webpage, undated-b. As of January 25, 2021:
https://gis.cdc.gov/Cancer/USCS/DataViz.html

———, "Tuberculosis (TB): Treatment for TB Disease," webpage, last reviewed April 5, 2016. As of January 25, 2021:
https://www.cdc.gov/tb/topic/treatment/tbdisease.htm

———, *National Diabetes Statistics Report 2020: Estimates of Diabetes and Its Burden in the United States*, Atlanta, Ga., 2020a. As of January 25, 2021:
https://www.cdc.gov/diabetes/pdfs/data/statistics/national-diabetes-statistics-report.pdf

———, "TB Incidence in the United States, 1953–2019," webpage, last reviewed October 24, 2020b. As of January 21, 2020:
https://www.cdc.gov/tb/statistics/tbcases.htm

Cetus Corporation, *Cetus Corporation Annual Report*, 1990. As of January 28, 2021:
https://quod.lib.umich.edu/c/cohenaids/5571095.0593.005?rgn=main;view=fulltext

Chang, Annie C. Y., and Stanley N. Cohen, "Genome Construction Between Bacterial Species *In Vitro*: Replication and Expression of *Staphylococcus* Plasmid Genes in *Escherichia coli*," *Proceedings of the National Academy of Sciences of the United States of America*, Vol. 71, No. 4, 1974, pp. 1030–1034.

———, "Construction and Characterization of Amplifiable Multicopy DNA Cloning Vehicles Derived from the P15A Cryptic Miniplasmid," *Journal of Bacteriology*, Vol. 134, No. 3, 1978, pp. 1141–1156.

Chetty, Raj, John N. Friedman, Nathaniel Hilger, Emmanuel Saez, Diane Whitmore Schanzenbach, and Danny Yagan, "How Does Your Kindergarten Classroom Affect Your Earnings? Evidence from Project STAR," *Quarterly Journal of Economics*, Vol. 126, No. 4, 2011, pp. 1593–1660.

Clarivate, "Editorial Selection Process," webpage, undated-a. As of January 21, 2021:
https://clarivate.com/webofsciencegroup/solutions/editorial

———, "KeyWords Plus Generation, Creation, and Changes," webpage, undated-b. As of January 21, 2021:
https://support.clarivate.com/ScientificandAcademicResearch/s/article/KeyWords-Plus-generation-creation-and-changes?language=en_US

———, "Web of Science Core Collection," webpage, undated-c. As of January 21, 2021:
https://clarivate.com/webofsciencegroup/solutions/web-of-science-core-collection/

Clark-Lewis, Ian, Leroy E. Hood, and Stephen B. Kent, "Role of Disulfide Bridges in Determining the Biological Activity of Interleukin 3," *Proceedings of the National Academy of Sciences of the United States of America*, Vol. 85, No. 21, 1988, pp. 7897–7901.

Cohen, Stanley N., "Comparison of Autologous, Homologous, and Heterologous Normal Skin Grafts in the Hamster Cheek Pouch," *Experimental Biology and Medicine*, Vol. 106, No. 4, 1961, pp. 677–680.

———, "Science, Biotechnology, and Recombinant DNA: A Personal History," an oral history conducted by Sally Smith Hughes in 1995, Regional Oral History Office, The Bancroft Library, University of California, Berkeley, 2009.

Cohen, Stanley N., Annie C. Y. Chang, Herbert W. Boyer, and Robert B. Helling, "Construction of a Biologically Bacterial Plasmids *In Vitro*," *Proceedings of the National Academy of Sciences of the United States of America*, Vol. 70, No. 11, 1973, pp. 3240–3244.

Cook, Lisa D., "Violence and Economic Activity: Evidence from African American Patents, 1870–1940," *Journal of Economic Growth*, Vol. 19, 2014, pp. 221–257.

Danna, Kathleen, and Daniel Nathans, "Specific Cleavage of Simian Virus 40 DNA by Restriction Endonuclease of Hemophilus Influenzae," *Proceedings of the National Academy of Sciences of the United States of America*, Vol. 68, No. 12, 1971, pp. 2913–2917.

Derynck, Rik, Julie A. Jarrett, Ellson Y. Chen, Dennis H. Eaton, John R. Bell, Richard K. Assoian, Anita B. Roberts, Michael B. Sporn, and David V. Goeddel, "Human Transforming Growth Factor-β Complementary DNA Sequence and Expression in Normal and Transformed Cells," *Nature*, Vol. 316, 1985, pp. 701–705.

DeSimone, Joseph, transcript of an interview by David J. Caruso and Jody A. Roberts at University of North Carolina, Chapel Hill, North Carolina, May 1–2, 2013. As of January 13, 2021:
https://oh.sciencehistory.org/sites/default/files/desimone_jm_0701_suppl.pdf

Devaraj, Neal K., "The Future of Bioorthogonal Chemistry," *ACS Central Science*, Vol. 4, No. 8, 2018, pp. 952–959. As of January 26, 2021:
https://pubs.acs.org/doi/10.1021/acscentsci.8b00251#

Diamond v. Chakrabarty, 447 U.S. 303, 1980.

Drake, Penelope, "Catalent's SMARTag® ADC Stands Out in the Crowd of HER2-Targeted Conjugates," *ADCReview*, September 29, 2020. As of November 12, 2020:
https://www.adcreview.com/must-read-articles/catalents-smartag-adc-stands-out-in-the-crowd-of-her2-targeted-conjugates

Duca, Lindsey M., Bing Wang, Marian Rewers, and Arleta Rewers, "Diabetic Ketoacidosis at Diagnosis of Type 1 Diabetes Predicts Poor Long-Term Glycemic Control," *Diabetes Care*, Vol. 40, No. 9, 2017, pp. 1249–1255.

Edelson, Steve, "Versant Ventures Launches Lycia Therapeutics with $50 Million," *Business Wire*, press release, June 9, 2020. As of January 25, 2021:
https://www.businesswire.com/news/home/20200609005070/en/Versant-Ventures-Launches-Lycia-Therapeutics-50-Million

Enable Biosciences, "ADAP SARS-CoV-2 Total Antibody Assay," webpage, undated-a. As of January 28, 2021. As of January 28, 2021:
https://www.enablebiosciences.com/serum

————, "Product Pipeline," webpage, undated-b. As of January 28, 2021:
https://www.enablebiosciences.com/pipeline

Epogen, homepage, undated. As of January 28, 2021:
https://www.epogen.com

EY, *Biotechnology Report 2017: Beyond Borders, Staying the Course*, 2017. As of December 31, 2020:
https://assets.ey.com/content/dam/ey-sites/ey-com/en_gl/topics/life-sciences/life-sciences-pdfs/ey-biotechnology-report-2017-beyond-borders-staying-the-course1.pdf

Fagerberg, Jan, "Innovation: A Guide to the Literature," in Jan Fagerberg and David C. Mowery, eds., *Innovation: A Guide to the Literature*, Oxford, U.K.: Oxford University Press, September 2009. As of January 26, 2021:
https://www.oxfordhandbooks.com/view/10.1093/oxfordhb/9780199286805.001.0001/oxfordhb-9780199286805-e-1

Fang, Yinzhi, Han Zhang, Zhen Huang, Samuel L. Scinto, Jeffery C. Yang, Christopher W. am Ende, Olga Dmitrenko, Douglas S. Johnson, and Joseph M. Fox, "Photochemical Syntheses, Transformations, and Bioorthogonal Chemistry of Trans-Cycloheptene and Sila Trans-Cycloheptene Ag(I) Complexes," *Chemical Science*, Vol. 9, 2018, pp. 1953–1963.

Fechner, Holly, and Matthew S. Shapanka, "Closing Diversity Gaps in Innovation: Gender, Race, and Income Disparities in Patenting and Commercialization of Inventions," *Technology & Innovation*, Vol. 19, 2018, pp. 727–734.

Feldman, Maryann P., Alessandra Colaianni, and Connie Kang Liu, "Lessons from the Commercialization of the Cohen-Boyer Patents: The Stanford University Licensing Program," in Anatole Krattiger, Richard T. Mahoney, Lita Nelsen, et al., eds., *Intellectual Property Management in Health and Agricultural Innovation: A Handbook of Best Practices*, Oxford, United Kingdom, Centre for the Management of Intellectual Property in Health and Development and Public Intellectual Property Resource for Agriculture, 2007, pp. 1797–1808.

Fliegel, L., K. Burns, D. H. MacLennan, R. A. Reithmeier, and M. Michalak, "Molecular Cloning of the High Affinity Calcium-Binding Protein (Calreticulin) of Skeletal Muscle Sarcoplasmic Reticulum," *Journal of Biological Chemistry*, Vol. 264, No. 36, 1989, pp. 21522–21528.

Flynn, Gregory C., Jan Pohl, Mark T. Flocco, and James E. Rothman, "Peptide-Binding Specificity of the Molecular Chaperone BiP." *Nature*, Vol. 353, 1991, pp. 726–730.

Garde, Damian, "Catalent Increases Stake in Redwood Bioscience" *Fierce Biotech*, March 31, 2014. As of November 12, 2020:
https://www.fiercebiotech.com/cro/catalent-increases-stake-redwood-bioscience

"The Gene Transplanters," *Newsweek*, June 17, 1974, p. 54.

Genentech, "Company Information," homepage, undated. As of December 10, 2020:
https://www.gene.com/media/company-information

⸻, Form 10-K Annual Reports, 1995–2008. Retrieved from U.S. Securities and Exchange Commission, Electronic Data Gathering, Analysis, and Retrieval System (EDGAR).

Genome News Network, *Genetics and Genomics Timeline: 1986,* webpage, undated. As of November 3, 2020:
http://www.genomenewsnetwork.org/resources/timeline/1986_Hood.php

"Getting Bacteria to Manufacture Genes," *San Francisco Chronicle*, May 21, 1974.

Gilbert, W., and D. Dressler, "DNA Replication: The Rolling Circle Model," *Cold Spring Harbor Symposia on Quantitative Biology*, Vol. 33, 1968, pp. 473–484.

Gilbert, Walter, and Allan Maxam, "The Nucleotide Sequence of the *lac* Operator," *Proceedings of the National Academy of Sciences of the United States of America*, Vol. 70, No. 12, 1973, pp. 3581–3584.

Girardie, Josiane, Adrien Girardie, Jean-Claude Huet, and Jean-Claude Pernollet, "Amino Acid Sequence of Locust Neuroparsins," *FEBS Letters*, Vol. 245, Nos. 1–2, 1989, pp. 4–8.

Google Scholar, "Google Scholar Metrics," webpage, undated. As of January 19, 2021:
https://scholar.google.com/intl/en/scholar/metrics.html#metrics

Gould, Michael K., Tania Tang, In-Lu Amy Liu, Janet Lee, Chengyi Zheng, Kim. N. Danforth, Anne E. Kosco, Jamie L. Di Fiore, and David E. Suh, "Recent Trends in the Identification of Incidental Pulmonary Nodules," *American Journal of Respiratory and Critical Care Medicine*, Vol. 192, No. 10, 2015, pp.1208–1214.

Griffin, B. A., S. R. Adams, and R. Y. Tsien, "Specific Covalent Labeling of Recombinant Protein Molecules Inside Live Cells," *Science*, Vol. 281, No. 5374, 1998, pp. 269–272.

Grosselin, Peter G., and Paul Jacobs, "DNA Device's Heredity Scrutinized by U.S.," *Los Angeles Times*, May 14, 2000. As of November 12, 2020:
https://www.latimes.com/archives/la-xpm-2000-may-14-mn-30009-story.html

Gugliotta, Guy, "Deciphering Old Texts, One Woozy, Curvy Word at a Time," *New York Times,* March 28, 2011.

Hall, Bronwyn H., Adam Jaffe, and Manuel Trajtenberg, "Market Value and Patent Citations," *RAND Journal of Economics*, Vol. 36, No. 1, 2005, pp. 16–38.

Hall, Bronwyn, Jacques Mairesse, and Pierre Mohnen, "Measuring the Returns to R&D, National Bureau of Economic Research," National Bureau of Economic Research Working

Paper 15622, December 2009. As of November 10, 2020:
https://www.nber.org/system/files/working_papers/w15622/w15622.pdf

Handyside, A. H., J. K. Pattinson, R. J. Penketh, J. D. Delhanty, R. M. Winston, and E. G. Tuddenham, "Biopsy of Human Preimplantation Embryos and Sexing by DNA Amplification," *The Lancet*, Vol. 333, No. 8634, 1989, pp. 347–349.

Hang, Howard C.; Chung Yu, Darryl L. Kato, and Carolyn R. Bertozzi, "A Metabolic Labeling Approach Towards Proteomic Analysis of Mucin-type O-linked Glycosylation," *Proceedings of the National Academy of Sciences of the United States of America*, Vol. 100, No. 25, 2003, pp. 14846–14851.

Healy, Mark, and Martin Flores, "Captcha If You Can," *Ceros*, undated. As of November 11, 2020:
https://www.ceros.com/originals/recaptcha-waymo-future-of-self-driving-cars/

Hewick, R. M., M. W. Hunkapiller, L. E. Hood, and W. J. Dreyer, "A Gas-Liquid Solid Phase Peptide and Protein Sequenator," *Journal of Biological Chemistry*, Vol. 256, No. 15, 1981, pp. 7990–7997.

Heyneker, Herbert L., John Shine, Howard M. Goodman, Herbert W. Boyer, John Rosenberg, Richard E. Dickerson, Saran A. Narang, Keiichi Itakura, Sry-yaung Lin, and Arthur D. Riggs, "Synthetic *lac*operator DNA is Functional *in Vivo*," *Nature*, Vol. 263, 1976, pp. 748–752.

Higuchi, Russell, Cecilia H. von Beroldingen, George F. Sensabaugh, and Henry A. Erlich, "DNA Typing from Single Hairs," *Nature*, Vol. 332, No. 6164, 1988, pp. 543–545.

Hood, Leroy Edward, *Immunoglobulins: Structure, Genetics and Evolution*, dissertation, Pasadena, Calif.: California Institute of Technology, 1968.

———— "A Personal View of Molecular Technology and How It Has Changed Biology," *Journal of Proteome Research*, Vol. 1, No. 5, 2002, pp. 399–409.

————, "A Personal Journey of Discovery: Developing Technology and Changing Biology," *Annual Review of Analytic Chemistry*, Vol. 1, 2008, pp. 1–43.

Horton, Robert M., Zeling Cai, Steffan N. Ho, and Larry R. Pease, "Gene Splicing by Overlap Extension: Tailor-Made Genes Using the Polymerase Chain Reaction," *Biotechniques*, Vol. 8, No. 5, 1990, pp. 528–535.

Horton, Robert M., Henry D. Hunt, Steffan N. Ho, Jeffrey K. Pullen, and Larry R. Pease, "Engineering Hybrid Genes Without the Use of Restriction Enzymes: Gene Splicing by Overlap Extension," *Gene*, Vol. 77, No. 1, 1989, pp. 61–68.

Horvath, Suzanna J., Joseph R. Firca, Tim Hunkapiller, Michael W. Hunkapiller, and Leroy Hood, "An Automated DNA Synthesizer Employing Deoxynucleoside 3′-Phosphoramidites," *Methods in Enzymology*, Vol. 154, 1987, pp. 314–326.

Hughes, Sally Smith, *Genentech: The Beginnings of Biotech,* Chicago: University of Chicago Press, 2011.

"Human Insulin Market Size, Share & Industry Analysis, By Type (Analogue Insulin, Traditional Human Insulin), By Diabetes Type (Type 1, Type 2), By Distribution Channel (Retail Pharmacy, Hospital Pharmacy, Online Pharmacy), and Regional Forecast, 2019–2026," *Fortune Business Insights*, webpage, January 2020. As of December 31, 2020: https://www.fortunebusinessinsights.com/industry-reports/human-insulin-market-100395

Iancu, Andrei, and Laura A. Peter, *Report to Congress Pursuant to P.L. 115-273, the SUCCESS Act*, Washington, D.C.: U.S. Patent and Trademark Office, October 2019. As of January 13, 2020, available at https://www.uspto.gov/sites/default/files/documents/USPTOSuccessAct.pdf#:~:text=The%20Study%20of%20Underrepresented%20Classes%20Chasing%20Engineering%20and,and%20veterans%20and%20the%20small%20businesses%20they%20own

IBISWorld, *Biotechnology in the US,* 2020. Subscription required.

Inouye, Satoshi, Masato Noguchi, Yoshiyuki Sakaki, Yasuyuki Takagi, Toshiyuki Miyata, Sadaaki Iwanaga, and Frederick I. Tsuji, "Cloning and Sequence Analysis of cDNA for the Luminescent Protein Aequorin," *Proceedings of the National Academy of Sciences of the United States of America*, Vol. 82, No. 10, 1985, pp. 3154–3158.

Institute for Systems Biology, "2018 Annual Report," webpage, undated-a. As of January 26, 2021: https://isbscience.org/annualreport/2018

———, "About," webpage, undated-b. As of January 29, 2021: https://isbscience.org/about/overview/

———, "Brain Health," webpage, undated-c. As of January 29, 2021: https://isbscience.org/research/brain-health

International Human Genome Sequencing Consortium, "Initial Sequencing and Analysis of the Human Genome," *Nature*, Vol. 409, 2001, pp. 860–921.

———, "Finishing the Euchromatic Sequence of the Human Genome," *Nature*, Vol. 431, 2004, pp. 931–945.

IPVision, dataset, undated. Not available to the general public.

ISB—*See* Institute for System Biology.

Itakura, K., T. Hirose, R. Crea, A. Riggs, H. L. Heyneker, F. Bolivar, and H. W. Boyer, "Expression in Escherichia Coli of a Chemically Synthesized Gene for the Hormone Somatostatin," *Science*, Vol. 198, No. 4321, 1977, pp. 1056–1063.

Jacobson, Lea, "Thios Pharmaceuticals, Inc." *Start Up*, October 1, 2002. As of November 12, 2020:
https://scrip.pharmaintelligence.informa.com/SC090661/Thios-Pharmaceuticals-Inc

JDRF, "The Complexity of Diagnosing Type 1 Diabetes," webpage, undated. As of January 25, 2021:
https://www.jdrf.org/t1d-resources/about/diagnosis

Jiang, Hao, Lei Feng, David Soriano del Amo, Ronald D. Seidel III, Florence Marlow, and Peng Wu, "Imaging Glycans in Zebrafish Embryos by Metabolic Labeling and Bioorthogonal Click Chemistry," *Journal of Visualized Experiments,* Vol. 52, 2011.

Kamariza, Mireille, Peyton Shieh, Christopher S. Ealand, Julian S. Peters, Brian Chu, Frances P. Rodriguez-Rivera, Mohammed R. Babu Sait, William V. Treuren, Neil Martinson, Rainer Kalscheuer, Bavesh D. Kana, and Carolyn R. Bertozzi, "Rapid Detection of Mycobacterium Tuberculosis in Sputum with a Solvatochromic Trehalose Probe," *Science Translational Medicine*, Vol. 10, No. 430, 2018.

Kelly, Bryan, Dimitris Papanikolaou, Amit Seru, and Matt Taddy, "Measuring Technological Innovation over the Long Run," National Bureau of Economic Research Working Paper No. 25266, November 2018, revised in February 2020.

Kelsey, Frances O., "Problems Raised for the FDA by the Occurrence of Thalidomide Embryopathy in Germany, 1960–1961," *American Journal of Public Health and the Nation's Health*, Vol. 55, No. 5, 1965, pp. 703–707.

Khan, Suliman, Muhammad Wajid Ullah, Rabeea Siddique, Gulam Nabi, Sehrish Manan, Muhammad Yousaf, and Hongwei Hou, "Role of Recombinant DNA Technology to Improve Life," *International Journal of Genomics*, 2016.

Kidd, Kimiko, "Here's Why CAPTCHA Shows You Traffic Pictures," *The News Wheel*, March 19, 2019. As of November 11, 2020:
https://thenewswheel.com/captcha-self-driving-cars

Knight, E. Jr., M. W. Hunkapiller, B. D. Korant, R. W. Hardy, and L. E. Hood, "Human Fibroblast Interferon: Amino Acid Analysis and Amino Terminal Amino Acid Sequence," *Science*, Vol. 207, No. 4430, 1980, pp. 525–526.

Kogan, Scott C., Marie Doherty, and Jane Gitschier, "An Improved Method for Prenatal Diagnosis of Genetic Diseases by Analysis of Amplified DNA Sequences," *New England Journal of Medicine*, Vol. 317, No. 16, 1987, pp. 985–990.

Kohmura, Masanori, Noriki Nio, and Yasuo Ariyoshi, "Solid-Phase Synthesis of [AsnA16]-, [AsnA22]-,[GlnA25]-, and [AsnA26]Monellin, Analogues of the Sweet Protein Monellin," *Bioscience, Biotechnology, and Biochemistry*, Vol. 56, No. 3, 1992, pp. 472–476.

Krieger, Lisa M., "Stanford Stem Cell Product, Delayed for More Than a Decade, to Be Tested Again," *Mercury News*, June 13, 2015, updated August 12, 2016. As of November 10, 2020: https://www.mercurynews.com/2015/06/13/stanford-stem-cell-product-delayed-for-more-than-a-decade-to-be-tested-again

Krischer, Jeremy, Kristian F. Lynch, Desmond A. Schatz, Jorma Ilonen, Åke Lernmark, William A. Hagiopian, Marian J. Rewers, Jin-Xiong She, Olli G. Simell, Jorma Toppari, Annete-G. Ziegler, Beena Akolkar, Ezio Bonifacio, and the TEDDY Study Group, "The 6 Year Incidence of Diabetes-Associated Autoantibodies in Genetically At-Risk Children: The TEDDY Study," *Diabetologia*, Vol. 58, No. 5, 2015, pp. 980–987.

Lampe, Ryan, "Strategic Citation," *Review of Economics and Statistics*, Vol. 94, No. 1, 2012, pp. 320–333.

Lansat, Myelle, and Richard Feloni, "The CEO Who Invented an Online Tool You See Every Day Gave His Tech to Yahoo for Free—And He Doesn't Regret It," *Business Insider*, June 26, 2018. As of January 28, 2021: https://www.businessinsider.com/duolingo-ceo-invented-captcha-gave-to-yahoo-for-free-2018-6

Laughlin, Scott T., Jeremy M. Baskin, Sharon L. Amacher, and Carolyn R. Bertozzi, "In Vivo Imaging of Membrane-Associated Glycans in Developing Zebrafish," *Science*, Vol. 320, No. 5876, 2008, pp. 664–667.

Lee, Tim, "Cerus Forms Group to Research Optimal Production of COVID-19 Convalescent Plasma," *Business Wire*, press release, March 26, 2020. As of January 25, 2021: https://www.businesswire.com/news/home/20200326005211/en/Cerus-Forms-Group-Research-Optimal-Production-COVID-19.

Lee, Victoria, and Andrew Grant, "Task Force on Artificial Intelligence—Hearing to Discuss Use of AI in Contact Tracing," *Technology's Legal Edge*, blog post, July 29, 2020. As of January 4, 2021: https://www.technologyslegaledge.com/2020/07/task-force-on-artificial-intelligence-hearing-to-discuss-use-of-ai-in-contact-tracing/

Lemelson Foundation, "Our Founders," webpage, undated. As of January 16, 2021: https://www.lemelson.org/our-story/our-founders

Lemelson-MIT, "$500k Prize," webpage, undated-a. As of January 16, 2021: https://lemelson.mit.edu/500k-prize-list

———, "Award Winners," webpage, undated-b. As of January 16, 2021:
https://lemelson-teams.mit.edu/winners_circle

———, "The Lemelson-MIT Student Prize for Collegiate Innovators," webpage, undated-c. As of January 16, 2021:
https://lemelson.mit.edu/studentprize

Linn, S. and W. Arber, "Host Specificity of DNA Produced by Escherichia Coli, X. In Vitro Restriction of Phage fd Replicative Form," *Proceedings of the National Academy of Sciences of the United States of America*, Vol. 59, No. 4, 1968, pp. 1300–1306.

Mamiya, Gunji, Kunio Takishima, Mayumi Masakuni, Tatsuko Kayumi, Kazuko Ogawa, and Tsunejiro Sekita, "Complete Amino Acid Sequence of Jack Bean Urease," *Proceedings of the Japan Academy*, Series B 61, No. 8, 1985, pp. 395–398.

Manyika, James, Susan Lund, Michael Chui, Jacques Bughin, Jonathan Woetzel, Parul Batra, Ryan Ko, and Saurabh Sanghvi, *Jobs Lost, Jobs Gained: What the Future of Work Will Mean for Jobs, Skills, and Wages*, McKinsey Global Institute, 2017. As of January 16, 2021:
https://www.mckinsey.com/~/media/McKinsey/Industries/Public%20and%20Social%20Sect or/Our%20Insights/What%20the%20future%20of%20work%20will%20mean%20for%20job s%20skills%20and%20wages/MGI-Jobs-Lost-Jobs-Gained-Report-December-6-2017.pdf

Matsuo, Kazuya, Yuki Nishikawa, Marie Masuda, and Itaru Hamachi, "Live-Cell Protein Sulfonylation Based on Proximity-Driven N-Sulfonyl Pyridone Chemistry," *Angewandte Chemie International Edition*, Vol. 57, No. 3, 2018, pp. 659–662.

Maxam, A. M., and W. Gilbert, "A New Method for Sequencing DNA," *Proceedings of the National Academy of Sciences of the United States of America*, Vol. 74, No. 2, 1977, pp. 560–564.

McCarthy, Alice A., "Thios Pharmaceuticals: Targeting Sulfation Pathways" *Chemistry & Biology*, Vol. 11, No. 2, 2004, pp. 147–148.

McElheny, Victor K., "Animal Gene Shifted to Bacteria; Aid Seen to Medicine and Farm," *New York Times*, May 20, 1974.

Meselson, Matthew, and Robert Yuan, "DNA Restriction Enzyme from E. Coli," *Nature*, Vol. 217, 1968, pp. 1110–1114.

Montague, P. Read, Carolyn D. Gancayco, Mark J. Winn, Richard B. Marchase, and Michael J. Friedlander, "Role of NO Production in NMDA Receptor-Mediated Neurotransmitter Release in Cerebral Cortex," *Science*, Vol. 263, No. 5149, 1994, pp. 973–977.

Morrow, J., S. Cohen, A. Chang, H. Boyer, H. Goodman, and R. Helling, "Replication and Transcription of Eukaryotic DNA in Escherichia Coli," *Proceedings of the National Academy of Sciences of the United States of America*, Vol. 71, No. 5, 1974, pp. 1743–1747.

Moser, Petra, and Tom Nicholas, "Prizes, Publicity and Patents: Non-Monetary Awards as a Mechanism to Encourage Innovation," *Journal of Industrial Economics*, Vol. 61, No. 3, 2013, pp. 763–788.

Moser, Petra, Joerg Ohmstedt, and Paul W. Rhode, "Patent Citations—An Analysis of Quality Differences and Citing Practices in Hybrid Corn," *Management Science*, Vol. 64, No. 4, 2018, pp. 1477–1973.

National Institutes of Health, National Cancer Institute, "Understanding Cancer: Cancer Statistics," webpage, last updated September 25, 2020. As of December 31, 2020: https://www.cancer.gov/about-cancer/understanding/statistics

National Institutes of Health, Recombinant DNA Advisory Committee, "Guidelines for Research Involving Recombinant DNA Molecules," 47 FR 167: 38048-68, August 27, 1982.

National Institutes of Health and National Human Genome Research Institute, "The Mouse Genome and the Measure of Man," *ScienceDaily*, December 5, 2002. As of November 10, 2020: https://www.sciencedaily.com/releases/2002/12/021205083819.htm

National Inventor's Hall of Fame, "Thomas J. Fogarty," webpage, undated. As of December 30, 2020: https://www.invent.org/inductees/thomas-j-fogarty

National Research Council, *Continuing Innovation in Information Technology*, Washington, D.C., The National Academies Press, 2012.

National Science Foundation, "Award Abstract (#9214821)Science & Technology Center for Molecular Biotechnology," webpage, undated. As of November 10, 2020: https://www.nsf.gov/awardsearch/showAward?AWD_ID=9214821

Nicholas, Tom, "Does Innovation Cause Stock Market Runups? Evidence from the Great Crash," *American Economic Review*, Vol. 98, No. 4, 2008, pp. 1370–1396.

Noel, Michael, and Mark Schankerman, "Strategic Patenting and Software Innovation," *Journal of Industrial Economics*, Vol. 61, No. 3, 2013, pp. 481–520.

Nokihara, Kiyoshi, and Toshihiko Semba, "Solid-Phase Synthesis of Porcine Cardiodilatin 88," *Journal of the American Chemical Society*, Vol. 110, No. 23, 1988, pp. 7847–7854.

Odasso, Cristina, Giuseppe Scellato, and Elisa Ughetto, "Selling Patents at Auction: An Empirical Analysis of Patent Value," *Industrial and Corporate Change*, Vol. 24, No. 2, 2015, pp. 417–438.

Office of Science and Technology Policy, "Coordinated Framework for Regulation of Biotechnology," *Federal Register*, Vol. 51, No. 123, June 26, 1986.

Olson, Maynard, Leroy Hood, Charles Cantor, and David Botstein, "A Common Language for Physical Mapping of the Human Genome," *Science*, Vol. 245, No. 4925, 1989, pp. 1434–1436.

Papadopoulos, Stelios, "Evolving Paradigms in Biotech IPO Valuations," *Nature Biotechnology*, Vol. 19, 2001, pp. BE18–BE19.

Parikka, V., K. Näntö-Salonen, M. Saarinen, T. Simell, J. Ilonen, H. Hyöty, R. Veijola, M. Knip, and O. Simell, "Early Seroconversion and Rapidly Increasing Autoantibody Concentrations Predict Prepubertal Manifestation of Type 1 Diabetes in Children at Genetic Risk," *Diabetologia*, Vol. 55, No. 7, 2012, pp. 1926–1936.

Park, Heon, Zhaoxia Li, Xuexian O. Yang, Seon Hee Chang, Roza Nurieva, Yi-Hong Wang, Ying Wang, Leroy Hood, Zhou Zhu, Qiang Tian, and Chen Dong, "A Distinct Lineage of CD4 T Cells Regulates Tissue Inflammation by Producing Interleukin 17," *Nature Immunology*, Vol. 6, No. 11, 2005, pp. 1133–1141.

Parraga, G., S. J. Horvath, A. Eisen, W. E. Taylor, L. Hood, E. T. Young, and R. E. Klevit, "Zinc-Dependent Structure of a Single-Finger Domain of Yeast ADR1," *Science*, Vol. 241, No. 4872, 1988, pp. 1489–1492.

PathCheck Foundation, "About," web page, undated. As of February 1, 2021: https://www.pathcheck.org/en/about

Perkin-Elmer, Form 10-K Annual Report, 1994. As of January 29, 2021: http://edgar.secdatabase.com/803/91335594000020/filing-main.htm

Philippidis, Alex, "Top 17 Serial Bio Entrepreneurs," *Genetic Engineering and Biotechnology News*, December 15, 2014. As of February 1, 2021: https://www.genengnews.com/a-lists/top-17-serial-bio-entrepreneurs

Pollack, Andrew, "Inventor of a DNA Sequencing Technique Is Disputed," *New York Times*, November 16, 2006.

Raftery, Michael A., Michael W. Hunkapiller, Catherine D. Strader, and Leroy E. Hood, "Acetylcholine Receptor: Complex of Homologous Subunits," *Science*, Vol. 208, No. 4451, 1980, pp. 1454–1456.

Rao, Justin M., and David H. Reiley, "The Economics of Spam," *Journal of Economic Perspectives*, Vol. 26, No. 3, 2012, pp. 87–110.

Rideout, Darryl, Theodora Calogeropoulou, James Jaworski, and Michael McCarthy, "Synergism Through Direct Covalent Bonding Between Agents: A Strategy for Rational Design of Chemotherapeutic Combinations," *Biopolymers*, Vol. 29, No. 1, 1990, pp. 247–262.

Riggins, Christie, "Great Inventions Don't Happen Overnight—Development of the Fogarty Balloon Catheter," *Stanford Medicine*, Vol. 17, No. 3, Fall 2000. As of January 16, 2021: http://sm.stanford.edu/archive/stanmed/2000fall/inventions.html

Roberts, Edward B., and Charles Eesley, *Entrepreneurial Impact: The Role of MIT*, Kansas City, Mo.: Ewing Marion Kauffman Foundation, 2009. As of December 22, 2020: https://www.kauffman.org/wp-content/uploads/2009/02/mit_impact_full_report.pdf

Roche, Finance Reports, 2009–2019. As of January 21, 2021: https://www.roche.com/investors/downloads.htm

Rogers, Mary A. M., Catherine Kim, Tanima Banerjee, and Joyce M. Lee, "Fluctuations in the Incidence of Type 1 Diabetes in the United States from 2001 to 2015: A Longitudinal Study," *BMC Medicine*, Vol. 15, No. 1, 2017.

Sarkis, Nichole, "Palleon Pharmaceuticals Raises $100 Million Series B to Develop Drugs Targeting Glycan-Mediated Immune Regulation to Treat Cancer and Inflammatory Diseases," Palleon Pharmaceuticals, press release, September 17, 2020. As of January 28, 2021: https://palleonpharma.com/press-releases/palleon-pharmaceuticals-raises-100-million-series-b-to-develop-drugs-targeting-glycan-mediated-immune-regulation-to-treat-cancer-and-inflammatory-diseases/

Schmid, Jon, *The Determinants of Military Technology Innovation and Diffusion*, dissertation, Atlanta: Georgia Institute of Technology, 2018.

Schmid, Jon, and Ayodeji Fajebe, "Variation in Patent Impact by Organization Type: An Investigation of Government, University, and Corporate Patents," *Science and Public Policy*, Vol. 46, No. 4, 2019, pp. 589–598.

Schmoch, Ulrich, *Concept of a Technology Classification for Country Comparisons: Final Report to the World Intellectual Property Organisation (WIPO)*, Karlsruhe, Germany: Fraunhofer Institute for Systems and Innovation Research, 2008.

Schnölzer, Martina, Paul Alewood, Alun Jones, Dianne Alewood, and Stephen B. H. Kent, "*In Situ* Neutralization in Boc-Chemistry Solid Phase Peptide Synthesis: Rapid, High Yield Assembly of Difficult Sequences," *International Journal of Peptide and Protein Research*, Vol. 40, Nos. 3–4, 1992, pp. 180–193.

Science History Institute, "Herbert W. Boyer and Stanley N. Cohen," webpage, last updated December 1, 2017. As of December 10, 2020: https://www.sciencehistory.org/historical-profile/herbert-w-boyer-and-stanley-n-cohen

Sharp, Phillip A., Bill Sugden, and Joe Sambrook, "Detection of Two Restriction Endonuclease Activities in Haemophilus Parainfluenzae Using Analytical Agarose-Ethidium Bromide Electrophoresis," *Biochemistry*, Vol. 12, No. 16, 1973, pp. 3055–3063.

Shreeve, James, *The Genome War: How Craig Venter Tried to Capture the Code of Life and Save the World*, New York: Random House Publishing Group, 2007.

Sinfield, Joseph, and Freddy Solis, "Finding a Lower-Risk Path to High-Impact Innovations," *MIT Sloan Management Review*, June 13, 2016. As of January 16, 2021: https://sloanreview.mit.edu/article/finding-a-lower-risk-path-to-high-impact-innovations

Sletten, Ellen M., and Carolyn R. Bertozzi, "From Mechanism to Mouse: A Tale of Two Biorthogonal Reactions," *Accounts of Chemical Research*, Vol. 44, No. 9, 2011, pp. 666–676. As of January 26, 2021: https://pubs.acs.org/doi/10.1021/ar200148z

Smith, H. O., and K. W. Wilcox, "A Restriction Enzyme from Haemophilus Influenzae. I. Purification and General Properties," *Journal of Molecular Biology*, Vol. 51, No. 2, 1970, pp. 379–391.

Smith, Lloyd M., Jane Z. Sanders, Robert J. Kaiser, Peter Hughes, Chris Dodd, Charles R. Connell, Cheryl Heiner, Stephen B. Kent, and Leroy E. Hood, "Fluorescence Detection in Automated DNA Sequence Analysis," *Nature*, Vol. 321, No. 6071, 1986, pp. 674–679.

"Solvay Prize 2020, Carolyn Bertozzi Wins for Bioorthogonal Chemistry," YouTube video, January 20, 2020. As of January 24, 2021: https://www.youtube.com/watch?v=YqGioIpJzlY

Steinberg, Fredric M., and Jack Raso, "Biotech Pharmaceuticals and Biotherapy: An Overview," *Journal of Pharmacy and Pharmaceutical Science*, Vol. 1, No. 2, 1998, pp. 48–59.

Sumner, James B., "The Isolation and Crystallization of the Enzyme Urease Preliminary Paper," *Journal of Biological Chemistry*, Vol. 69, No. 2, 1926, pp. 435–441.

Talanian, Robert V., C. James McKnight, Rheba Rutkowski, and Peter S. Kim, "Minimum Length of a Sequence-Specific DNA Binding Peptide," *Biochemistry*, Vol. 31, No. 30, 1992, pp. 6871–6875.

Temple, James, "Why Bad Things Happen to Clean-Energy Startups," *MIT Technology Review*, July 2017. As of January 16, 2021: https://www.technologyreview.com/2017/06/19/68258/why-bad-things-happen-to-clean-energy-startups

Thayer, Ann, "Biotech Firms Chiron and Cetus to Merge," *Chemical & Engineering News Archive*, Vol. 69, No. 30, 1991, p. 8.

"Thios Pharmaceuticals Inc.," *Biocentury*, webpage, undated. As of January 24, 2021:
https://bciq.biocentury.com/companies/thios_pharmaceuticals_inc

Thumm, Nikolaus, "Strategic Patenting in Biotechnology," *Technology Analysis & Strategic Management*, Vol. 16, No. 4, 2004, pp. 529–538.

Timmerman, Luke, *Hood: Trailblazer of the Genomics Age*, Bandera Press, 2016.

Tong, Amber, "Going Where Protacs Can't, Versant Unveils $50M Bet on Carolyn Bertozzi's LYTAC Tech—with a Rising Star at the Helm," *EndPoint News*, June 9, 2020. As of January 25, 2021:
https://endpts.com/going-where-protacs-cant-versant-unveils-50m-bet-on-carolyn-bertozzis-lytac-tech-with-a-seasoned-biotech-exec-at-the-helm/.

Toole, Andrew A., et al., *Progress and Potential: A Profile of Women Inventors on US Patents*, Washington, D.C.: U.S. Patent and Trademark Office, 2019.

Trefis Team and Great Speculations, "Was The $47 Billion Acquisition of Genentech in 2009 a Good Deal for Roche?" *Forbes*, July 10, 2019. As of October 9, 2020:
https://www.forbes.com/sites/greatspeculations/2019/07/10/was-the-47-billion-acquisition-of-genentech-in-2009-a-good-deal-for-roche/#6f77536ea1aa

Triphase Research and Development III Corporation, "Study of TRPH-222 in Patients with Relapsed and/or Refractory B-Cell Lymphoma," last updated January 20, 2020. As of January 25, 2021:
https://clinicaltrials.gov/ct2/show/NCT03682796

U.S. Census Bureau, "North American Industry Classification System," webpage, undated. As of January 16, 2021:
https://www.census.gov/eos/naics

U.S. Office of Management and Budget, Economic Classification Policy Committee, *North American Industry Classification System*, Washington, D.C., 2017.

U.S. Patent and Trademark Office, "General Information Concerning Patents," webpage, last updated June 1, 2020. As of January 19, 2021:
https://www.uspto.gov/patents/basics

U.S. Patent and Trademark Office, "Glossary," webpage, 2021. As of January 16, 2021:
https://www.uspto.gov/learning-and-resources/glossary

USPTO—*See* U.S. Patent Trademark Office.

U.S. Small Business Administration, "Table of Small Business Size Standards Matched to North American Industry Classification System Codes," effective August 19, 2019. As of June 1, 2020:

https://www.sba.gov/sites/default/files/2019-08/
SBA%20Table%20of%20Size%20Standards_Effective%20Aug%2019%2C%202019_
Rev.pdf

Van Dam, Andrew, "Step Aside Edison, Tesla And Bell. New Measurement Shows When U.S. Inventors Were Most Influential," *Washington Post*, November 27, 2018. As of September 9, 2020:
https://www.washingtonpost.com/business/2018/11/27/is-third-wave-us-innovation-over-what-we-learned-measurement-americas-most-influential-inventions/#comments-wrapper

Versant Ventures, "Lycia Therapeutics," webpage, undated. As of January 28, 2021:
https://www.versantventures.com/portfolio/lycia-therapeutics

Vuturo, Andrea, "InterVenn Biosciences Announces Positive Interim Clinical Trial Results and Appoints Biotech Veteran Klaus Lindpaintner, M.D. as Chief Scientific and Medical Officer," *Business Wire*, press release, November 19, 2019. As of January 25, 2021:
https://www.businesswire.com/news/home/20191118005886/en/InterVenn-Biosciences-Announces-Positive-Interim-Clinical-Trial%20as%20of%20January%204

———, "InterVenn Biosciences Uses Glycoproteomics to Identify Potential COVID-19 Susceptibility Markers," *Business Wire*, press release, October 13, 2020. As of January 25, 2021:
https://www.businesswire.com/news/home/20201013005716/en/

Wadman, Meredith, "Economic Return from Human Genome Project Grows," *Nature*, blog post, June 12, 2013. As of November 10, 2020:
https://www.nature.com/news/economic-return-from-human-genome-project-grows-1.13187

Westerich, Kim J., Karthik S. Chandrasekaran, Theresa Gross-Thebing, Nadine Kueck, Erez Raz, and Andrea Rentmeister, "Bioorthogonal mRNA Labeling at the Poly (A) Tail for Imaging Localization and Dynamics in Live Zebrafish Embryos," *Chemical Science*, Vol. 11, 2020, pp. 3089–3095.

World Health Organization, "Tuberculosis," webpage, October 14, 2020. As of November 12, 2020:
https://www.who.int/news-room/fact-sheets/detail/tuberculosis

World Intellectual Property Organization, "Frequently Asked Questions: Patents," webpage, undated. As of January 16, 2020:
https://www.wipo.int/patents/en/faq_patents.html

Yang, Jiwei, Steven D. Rosen, Philip Bendele, and Stefan Hemmerich, "Induction of PNAd and N-Acetylglucosamine 6-O-Sulfotransferases 1 and 2 in Mouse Collagen-Induced Arthritis," *BMC Immunology*, Vol. 7, No. 1, 2006.

Yoshimori, Robert, Daisy Roulland-Dussoix, and Herbert W. Boyer, "R Factor-Controlled Restriction and Modification of Deoxyribonucleic Acid: Restriction Mutants," *Journal of Bacteriology*, Vol. 112, No. 3, 1972, pp. 1275–1279.

Yu, Jian, Da Shen, Hanjie Zhang, and Zheng Yin, "Rapid, Stoichiometric, Site-Specific Modification of Aldehyde-Containing Proteins Using a Tandem Knoevenagel-Intra Michael Addition Reaction," *Bioconjugate Chemistry*, Vol. 29, No. 4, 2018, pp. 1016–1020.

Ziegler, A.-G., E. Bonifacio, and BABYDIAB-BABYDIET Study Group, "Age-Related Islet Autoantibody Incidence in Offspring of Patients with Type 1 Diabetes," *Diabetologia*, Vol. 55, 2012, pp. 1937–1943.

Ziegler, Anette G., Marian Rewers, Olli Simell, Tuula Simell, Johanna Lempainen, Andrea Steck, Christiane Winkler, Jorma Ilonen, Riitta Veijola, Mikael Knip, Ezio Bonifacio, and George S. Eisenbarth, "Seroconversion to Multiple Islet Autoantibodies and Risk of Progression to Diabetes in Children," *JAMA*, Vol. 309, No. 23, 2013, pp. 2473–2479.

Zoon, Kathryn C., Mark E. Smith, Pamela J. Bridgen, Christian B. Anfinsen, Michael W. Hunkapiller, and Leroy E. Hood, "Amino Terminal Sequence of the Major Component of Human Lymphoblastoid Interferon," *Science*, Vol. 207, No. 4430, 1980, pp. 527–528.